D0265044

05286940

Breaking
&
Mending

Breaking
&
Mending

JOANNA
CANNON

P

PROFILE BOOKS

wellcome
collection

First published in Great Britain in 2019 by
PROFILE BOOKS LTD
3 Holford Yard
Bevin Way
London
WC1X 9HD
www.profilebooks.com

Published in association with Wellcome Collection

**wellcome
collection**

183 Euston Road
London NW1 2BE
www.wellcomecollection.org

1 3 5 7 9 10 8 6 4 2

Typeset in Garamond by MacGuru Ltd
Printed and bound in Great Britain by Clays Ltd, Elcograf S.p.A.

A CIP catalogue record for this book is
available from the British Library.

ISBN 978 1 78816 057 5
eISBN 978 1 78283 452 6

FSC
www.fsc.org
MIX
Paper from
responsible sources
FSC® C018072

For Michaela

Contents

Breaking

We always strive to do the best for our patients, to provide the best possible care – why do we not do the same for our colleagues?
The junior doctor

A few years ago, I found myself in A&E.

I had never felt so ill. I was mentally and physically broken. So fractured, in fact, that I hadn't eaten properly or slept well, or even changed my expression, for months. My hands shook. My eyes swam with too much seeing, and I sat in a cubicle, behind paper-thin curtains, listening to the rest of the hospital happen around me. It was the effort of not crying that stole the most energy. It felt as though the frame of my existence – the fragile scaffolding that held me together – was beginning to snap and splinter, and if no one reached out to help, if no one noticed, the very sense of who I was would soon be spilled from me and lost forever. I knew I was an inch away from defeat, from the acceptance of a failure I assumed would be inevitable, but equally, I knew I had to carry on. I had to somehow walk through it.

Because I wasn't the patient. I was the doctor.

Each time we visit a hospital, we see them. An army of scrubs and stethoscopes, travelling the corridors with a quiet confidence. We imagine, strangely, that they are invincible. That understanding the mechanism of a disease somehow prevents a person from contracting it. You will never find a cardiologist with angina, a respiratory physician cannot suffer with asthma, and a psychiatrist can never truly understand what it's like to live with depression. Fallacies. All of them. But perhaps necessary fallacies because they help us to hold on to the belief that a doctor has the power to save us – and if doctors are unable to save themselves, what hope can be offered to anyone else?

For some, though, a stethoscope is less of a protective talisman, and more of a risk factor, because it carries with it an unimaginable burden. The burden of what it *means* to be a doctor – the internal and external pressure of a definition we have shaped and polished since childhood. A definition moulded by films and television shows, by books and soap operas and magazines. Doctors are objective, calm, knowledgeable. Doctors protect and heal and mend. Doctors *fix things*. You spend five years at medical school, learning how to fix things, only to arrive on the wards and discover very quickly that there are many, many things you will never be able to fix. During those five years, you sit in front of endless exam papers – exam papers with empty white boxes waiting to be filled with the answers to what you would *do* in imaginary scenarios – only to find that in many real-life situations, the very best thing you can do is absolutely nothing at all. As the textbooks transform into living, breathing people and the imaginary scenarios become a reality, you will eventually learn that being a good doctor really has nothing whatsoever to do with fixing people. You will also learn that a failure to mend doesn't make you a failure, and you will learn that an empty white box is sometimes the correct answer after all. But you will only learn these things after walking hundreds of miles of hospital corridors, navigating a landscape so alien and so challenging, you wonder why you ever chose to walk through it in the first place.

But you will learn. Eventually. As long as the landscape doesn't break you first.

Each time I read that another doctor has vanished from their life, that someone else has felt the need to disappear from this landscape, it takes my breath away for a moment, because it could have been any one of us. It most definitely could have been me, as I sat there in an A&E cubicle trying to work out how a job I had been so determined to do, and so desperate to be good at, had turned itself into my nemesis. I thought back to medical school, to all the past moments that had tied and knotted together and had led me to this one. I thought even further back, to my medical school interview, when I spoke with such passion about a profession I wanted so badly to be a part of. This was my dream, my ultimate goal, and yet it had turned into a nightmare so vivid and so brutal that I could hardly bear to look any more.

On that day, in the middle of a busy emergency department, as I tried to claw my way back from the edge of a cliff, if you had shown me a door marked 'escape', I would have gladly walked through it.

Stories

Before medical school, the idea of becoming a doctor was nothing but a faraway dream. A dream based on those early experiences of my childhood, my memories of our general practitioner, a family doctor, of the operation and rehabilitation for my club foot, of my appendectomy. Those brief but intense moments that found their way into my chronological life and somehow carved themselves deeper than others. That carving formed a draft image of what a doctor was. My memory of feeling safe among the fear, of great abilities, but mainly of kindness. Those memories were of the doctor I wanted to be.

The consultant

At your medical school interview, among the many topics you might be encouraged to speak about, there is only one question you know will definitely be asked, only one question you can actually prepare for:

Tell me, why do you want to become a doctor?

We all say *'I want to become a doctor because I like people'*, but what we're really saying is that we like stories. Stories bind us together, stories unite us, and we tell our stories in the hope that someone out there will listen, and we will be understood.

Months later, those of us fortunate enough to be offered a place from that interview were thrown together on a grey September morning at the very beginning of medical school. We didn't realise that the experiences we were destined to share would bind us all for a lifetime, that walking together through the next five years would grow friendships, relationships – even marriages and children – because at that moment we were still strangers, held in the darkness of a lecture theatre, by a magical and breathless excitement.

It was our inaugural lecture. A learned and dazzlingly qualified professor stood on the stage before us, and he leaned on the podium and gazed out at his audience, in a learned and dazzlingly qualified silence, and we waited. All three hundred of us. When he finally spoke, he dug into the bones of how each of us felt. How we had felt in all the weeks beforehand, buying books from a list four pages long and staring endlessly at the timetable of our future selves. Being paraded in front of friends and relatives. Imagining. Daring to believe, but dismissing those beliefs as foolishness. We had all felt it that morning, as we walked or cycled, or drove to our new beginning. For every seat in that lecture theatre, there had been four other people who had wanted to sit there. Surely this meant we were capable? Surely this meant we were finally allowed to turn to long-held dreams and taste the possibility? Yet still we felt foolish, because it all seemed so ridiculous. So *unlikely*. But with his words on that grey September morning, a wise professor managed to find the very pinpoint of how each of us was feeling, and in that moment it all became less foolish. Less unlikely. In that moment, it all became real.

'Welcome,' he said, 'to the first day of your medical career.'

We were a mixed bunch of three hundred in that lecture theatre. There were some who came from family trees littered with doctors, some who were the first to see the inside of a university. There were those who had travelled only a few miles to be there, and those who had travelled halfway around the world. Some were fresh from A-levels or a gap year spent wandering the planet, while others – like me – had found medicine later in life, in their thirties or even in their forties, having arrived from a succession of seemingly unrelated jobs that would – in time – prove strangely useful. But it was a love of stories that made our common ground, and we would spend the rest of our lives listening to them. Stories told in the handkerchief quiet of palliative care, stories told in the rush of an outpatient clinic. Stories whispered under the crash and chaos of an emergency department. Funny stories. Sad stories. Stories woven with lies that needed to be unpicked. Stories that made us laugh, or despair or worry. Stories that made us smile on the drive home, and stories so profoundly moving, we would carry them with us for the rest of our lives.

I am very often asked about the similarities between being a doctor and being an author, and the answer is very simple. Writing always rests on a narrative, on hearing a voice, and it's exactly the same for medicine – because medicine is all about people, and people are made out of stories.

Wild Cards

Five years sounds like a long time to be studying for a degree, but in reality, it's the briefest of moments. Five short years in which to turn perfectly predicted grades into doctors. Not only to impart vast amounts of knowledge and information, but to impart a different way of looking at the world. A different attitude, a different identity. Some wear the identity with ease, but struggle with the workload. Others pass exams without difficulty, but discover their new identity is not one with which they feel comfortable. We have five years to rectify that, to nurture and support, to prepare and tidy. At the end of those five years, we have to let someone go and hope that we have done enough.

Sometimes, we haven't. Sometimes, they fracture. You select the highest achievers, the perfectionists, the school prefects and the sports captains, children who have spent the whole of their short lives being the best, being prize winners and medallists, being applauded and noticed and being accustomed to standing out. If you take those people and put them in a room with three hundred identical people, until no one stands out any more and those who have previously achieved A-grades with ease now have to fight and scramble just to keep up, and you add to that mix an overwhelming workload and an intense amount of pressure, it's understandable some of them break. In all honesty, I'm surprised it doesn't happen more often.

I remember all my students, but it's the ones who break that I remember most clearly, because I always wonder, if I had looked more closely, if I had concentrated a little better, perhaps I could have spotted the fracture lines in time. Perhaps I could have prevented it from happening.

The admissions tutor

I was a wild card.

An elderly and about-to-retire professor interviewed me for medical school, and the only other time I saw him – ironically – was graduation day. I thanked him for giving me a chance, not thinking for a second that he would remember me.

He did.

'Each year, I would pick an outsider. A high risk. That year, I picked you,' he said. 'You were my wild card.'

As wild cards go, I was pretty wild.

I had left school at fifteen with only one O-level and very little else. Like many children then and many children now, we are asked to make huge decisions about what we want to do with our lives before we've even really discovered who we are. At fifteen I had no idea, and so I left. I decided to think about it. I ended up thinking about it for quite a while.

I did lots of other jobs as I thought. I typed letters and pulled pints, and delivered pizzas. I worked at a wonderful animal rescue centre. I waited tables. I was one of those really annoying women in department stores, the ones who try to spray you with fragrance as you walk through the shop. The ones you try desperately to avoid. I was that woman, and I spent months of my life watching people run away from me.

But I never lost a vague and quiet hope at the back of my

mind that one day I would return to education. I never let go of my need to learn and I would do lots of things to satisfy that need. I would read textbooks just for fun. I watched documentaries on rare and unimaginable diseases. I did courses and workshops, and looked out for any other exposure to education, however small, that I could.

One August morning, in the summer of 2003, I saw a very basic first-aid course advertised on a postcard in a newsagent's window. I happened to look up as I walked through the door. A chance. A small moment tied and knotted to many other small moments, that eventually joined together and led me to being a doctor. I rang and booked myself a place, and in the coffee break on that course, I told the paramedic who was teaching us how much I loved medicine and how interested I was in psychiatry, but how – in my thirties – I was too old to even consider it now. He told me I wasn't. He told me people in their thirties, people in their forties, apply for medical school, and in a moment of wild spontaneity the very next day I enrolled to do three science A-levels. Just over a year later, I found myself sitting opposite the elderly and about-to-be retired professor in an interview room deep within the bowels of a medical school. He was concerned about my age.

'I'm worried how you'll cope with the workload at your age,' he said.

'I'm worried how you'll support yourself,' he said.

'I'm worried how you'll feel when the consultant you are working for is younger than you are,' he said.

I brushed these remarks, and many others, away as best I could, even the last one, which made me hesitate just for a moment.

The professor leaned back and folded his arms. He stared

at me in silence. I stared back. There were no more questions to answer and I decided I had nothing to lose.

'Look,' I said. 'I completely understand if you reject me. Reject me because you don't think I'm smart enough. Reject me because you don't think I'll make a very good doctor. Reject me for the hundred and one reasons you reject people, but please – please – don't reject me just because of my date of birth, because that wouldn't be a very good reason at all, would it?'

His eyebrows raised just ever so slightly. That's it, I thought, I've blown it.

A couple of weeks later, the offer of a place arrived in the post.

'Merry Christmas' it said, handwritten at the bottom of the letter.

I couldn't be certain, and it was never confirmed, but I think it might have been that little outburst of indignation that secured me my seat at medical school.

Hearts

There are many reasons why people decide to go to medical school, but if you had asked each of us on that first day why we were there, we would have told you it was because we wanted to make a difference. We would have told you it was because we wanted to do something valuable – something important. We would have told you it was because we wanted to save lives.

Saving lives is a big crowd-puller when it comes to medical school entry and you can truly understand it. Many years later, on my last placement before finals, I would find myself in the emergency department, the not-quite-a-doctor, trying desperately to avoid getting in anyone's way.

On one of my shifts, a woman was brought in by ambulance. A woman who was in her forties and usually in good health. A woman whose presenting complaint was a racing pulse and a feeling that something terrible was about to happen. A sense of impending doom. The general opinion was that she was having an anxiety attack (or rather, *just* having an anxiety attack, because society still likes to put 'just' in front of anything to do with mental health), and the woman sat in a cubicle in the assessment unit, and waited for a series of routine tests.

Ten minutes later, she went into cardiac arrest.

She slid, very gently, from the chair and on to the floor, and her heart stopped beating. If you have ever wondered about the definition of teamwork, you will find it on a crash call. It's a sharp, blinding algorithm of efficiency. A crash call has its own tribe, its own trolley, its own rules, and, as a medical student, I was told I should stay and watch. In the very next cubicle to the woman there just happened to be a consultant cardiologist seeing another patient, and he appeared from behind a curtain and took over.

And the cardiologist brought the woman back to life.

In a fusion of machinery and drugs and experience, her heart began to beat again. The cardiologist had managed to pull her away from wherever she had gone and bring her back into the world. She was resurrected. It was fast and clean. It was uncomplicated. The woman even tried to stand up (no, really). It was the first crash call I'd witnessed and I was mesmerised. I thought all crash calls would be like that (they are not). The woman was taken to somewhere more appropriate than a cubicle in an assessment unit and the floor was cleared of debris. The cardiologist turned to his audience and said 'she was right about the impending doom, wasn't she?' and he disappeared again behind a curtain – where I heard him apologise to his patient for leaving so abruptly, because cardiologists always seem to possess an impeccable sense of timing. The department carried on.

I did not, however, carry on. I was transfixed by what I had seen. I wanted to ask the cardiologist how it felt – how it felt to return a life to someone. How it felt to do a job where you could, at any moment, become a hero. How it felt to argue with God. But I didn't. I didn't ask him any of those things, because I had learned very quickly that in medicine and surgery, unless you enjoy being looked at in a very curious way, you do not ask people how something feels. Instead, my gaze followed him around A&E for the rest of the afternoon, and, whenever I spotted him, I thought: 'there is the man who saved a woman's life. There is the cardiologist. There is a hero.'

On that first day of medical school, if you had asked someone what their chosen specialty was going to be, cardiology would have been a very popular answer. 'It's the prestige,' people will tell you, because in medicine, there is a certain hierarchy of body parts, which I have never quite understood. When it comes to kudos, hearts trump brains, brains trump bones, bones trump skin. Kidneys would, of course, trump everything, but they're far too clever to involve themselves in such shenanigans. All I'd ever wanted to do was psychiatry (it was the whole reason I was sitting in that lecture theatre in the first place) although I would, in time, glance with awe at some of the other rotations as I passed through – the grace and compassion of palliative medicine, the utter joy of care of the elderly. But at the end of a very long road, I knew that psychiatry was waiting for me, and knowing that was sometimes the only thing that kept me going. As I travelled that road, though, I would sometimes remember the cardiologist and I would feel a whispering of regret that I would never know how it felt to kneel on the floor of an A&E department and to save someone's life.

It was only much later, when I finally reached the end of my journey, that I learned something vital. Perhaps the most vital thing you can ever learn as a junior doctor. I learned that saving a life often has nothing to do with a scalpel or a defibrillator. I learned that lives are not just saved on the floor of an A&E department or in a surgical theatre. Lives are also saved in quiet corners of a ward. During a conversation in a garden. On a sofa in a TV room, when everyone else has left. Lives can be saved by spotting something lying hidden in a history. Lives can be saved by building up so much trust with a patient, they will still take a medication even if they don't believe they need it. Lives can be saved by listening to someone who has spent their entire life never being heard.

I learned that returning a life to someone very often has nothing to do with restoring a heartbeat.

Bodies

Friendships from medical school remain intense despite the distance and our different stories. The friendship of those who were with me when I first saw a cadaver. Those who suffered with me from every illness we studied. Those who were with me when we first asked someone about their illness, or to undress in order to perform an examination, with a mixture of shyness and arrogance. This was our introduction to the privileged moments in people's lives that became our routines. These friendships helped us to navigate new encounters with pain and distress, and with joy. These encounters started to shape us as doctors in more ways than one, although we were still unaware of the dark corners of that new shape, still unaware that some of those friendships, and new ones, would eventually help us to shine light into those dark corners.

The consultant

I called them my Kodak moments. The small snippets of other people's lives that I took home with me every night. I collected many Kodak moments over the years and they filled album after album in my head. So many albums, in fact, that I soon began to wonder if I was cut from the right cloth to practise medicine after all.

If you walk a circuit of any hospital, you will find many of these Kodak moments in wards and clinics, and hidden behind the curtains of anonymous cubicles. If you wish to hunt them down, you will find plenty in intensive care and in A&E. Oncology usually has its fair share, and palliative medicine is swimming with them. Many Kodak moments, though, are found where you would least expect them to be – not held within the main story, but hidden at the edges of the narrative – because it's often the smaller moments, the incidental characters, that provide you with the memories that are the most difficult to leave behind. Whenever I expressed concern at how these moments affected me, I was always told that compassion is a wonderful thing. I was told compassion is something to be desired and applauded. But compassion will eat away at your sanity. It will make you pull up in a lay-by on the journey home, because you can no longer see the road for tears. It will creep through your mind in the darkness, and keep you from your sleep, and you will find that the cloth from which you are cut will begin to suffocate you.

It didn't take very long for the albums to begin filling. Exactly a week after sitting in a darkened lecture theatre and being welcomed into my medical career, I experienced my very first Kodak moment.

It was waiting for me under a cloth in the dissection room.

Anatomy, like many other things in life, is better learned by experience, rather than by reading about it. No matter how colourful and detailed the diagrams were in the giant textbooks we carried around, they were no match for seeing something in real life, and many of us had chosen our medical school purely because it conducted what is termed *full body dissection.* This means that you have a whole person to explore throughout your training. It is also the same person, the same cadaver, that you keep for your years as a medical student, and, as is tradition – and as many hundreds had experienced before us over the decades – we were introduced to 'our' cadaver in the very first week of medical school.

I knew it was coming. I had seen it creep towards me on the timetable. I felt prepared. Almost indifferent. I would be fine. It was animals that tore at my heartstrings, not people. I could deal with people. Although I was soon to discover that I actually couldn't deal with people very well at all.

Our first dissection was scheduled for the early afternoon and we gathered in the basement of the medical school, in our fresh white coats. Few of us had eaten any lunch. We drifted into clusters, small white knots of anxiety and appre-hension, bravado and curiosity. The armour of dark humour, with which I would soon become very familiar, began to creep around the edges of the room. I dug my hands deep into the pockets of my coat and tried to concentrate on the opportunity I was being given to learn, and the generosity of those who had donated their bodies in order to provide me with that opportunity. After what felt like a lifetime, we were ushered into the dissection room itself.

It was the aroma, more than anything. A unique blend of chemistry lab and death. The strange rubbery smell of preser-vative. More than that, though, it smelled of history and of tradition, because, as medical students, we were walking into an experience unchanged for almost three hundreds years, except that dissection was no longer held in giant auditori-ums, and graves were no longer robbed for the privilege.

Within the dissection room there were many tables, and upon those tables were clean blue sheets, and beneath those clean blue sheets lay dead bodies. We were split into groups and I stood with six others around our table. Our body. There then began a health and safety lecture, a new accessory to a three-hundred-year-old tradition, and as the words floated above my head I stared at the clean blue sheet and I wondered who might lie beneath it.

I thought of the last dead person I had seen, only a few months earlier. I had watched as my mother said goodbye to my father on a watery, pale, February morning, surrounded by the equipment you are loaned when someone dies at home. The hoists and the commode, and the bottles of Oramorph, the monitors and the Macmillan nurse, all crowded into your front room and trying so very hard to blend in with the furniture of an ordinary life.

You imagine when a doctor arrives at your bedside, or sits across a table from you in a consultation room, that they are somehow swept clean of their own reference points. The corners of their minds are tidied and orderly. They are unaffected by memory, or by difficult emotion, or by fracture lines caused by lives lived outside of that encounter. Another assumption. Another necessary fallacy. Because all I could think about, waiting in that dissection room and staring at the blue sheet, was my dad. I played out different scenarios in my mind as we were told about fire exits and suitable footwear. I thought about the consequences of leaving, and the potential outcomes of staying. I thought about how hard I had fought to be standing there and what other people might think of me if I walked away. Mainly, though, I thought about my dad. My mouth was dry. My pulse charged. My legs felt undecided about whether they wished to carry on with the job of holding me upright. I turned to the nearest person who looked as if they might be in charge of something and I explained. They listened and they understood. They told me to leave – and I left.

I staggered back through the medical school and into the sweet, fresh air of Lancaster Road, which did not smell of rubber or chemistry lab, and I sat in my car and I tried to find my breath again. I had failed. The first challenge medical school had offered up to me and I had snapped and broken. Even worse, it felt as though the dissection room had been an initiation, a ceremony almost. A rite of passage. While my future colleagues were still in that basement, transitioning into doctors, I was sitting in my car watching the rain hit the windscreen, and wondering why I had ever imagined I could do any of this in the first place.

Over the next two weeks, I tried many times to enter that dissection room. I ventured into the basement when no other students were around, thinking the solitude might help me to acclimatise to death. It didn't. I went down there to speak to one of the anatomists, perhaps looking for a little empathy, a little understanding. She explained to me the importance of dissection in the same language as the glossy pages in a medical school brochure, but I didn't listen because I was too busy staring just beyond her right shoulder, at a polythene bag containing twelve severed heads. I nodded and walked away. I even went to my GP, thinking I could medicate myself into facing up to it.

'I don't think I'm cut out for this,' I said. 'I don't think I can carry on.'

She stared at me. 'But you must. Especially now.'

I looked up at her from deep within my tearful, self-absorbed misery. 'Why?'

'Because the way you've reacted to the dissection room tells me what a good doctor you'll make.'

I didn't feel like I'd make a good doctor. I felt fraudulent. Ridiculous. As the dissection room rolled around each week on the timetable and as I rolled neatly away from it, I knew I had to either throw in the towel right there and then, or address it before it became insurmountable.

Along with the dissection room and genetics and physiology and pharmacology, and many other new and mysterious subjects, one of the topics we were presented with in our first semester at medical school was pathology. The registrar who taught us was around the same age as me, and she was funny and wise, and spoke with such passion and enthusiasm about her subject when she was with us that she made everyone in that room want to be a pathologist. She was also one of those rare people you couldn't help but instantly like, and in another of my moments of wild spontaneity, I decided to ask if I could accompany her to a post-mortem. Surely if I saw front-line anatomy in one of its most useful roles, solving puzzles and providing answers, it would help to rid me of my fears. But I was a first-year medical student. I had (quite literally) just walked in off the street. She would definitely say no, which was just as well, because if I couldn't face the clean, bloodless, preservation of a dissection room, how on earth would I have coped with an autopsy?

She said yes.

Choices

Whenever you visit a hospital, your destination is always clearly marked. Signposts are suspended above your head along all of the corridors. Arrows are painted on to floors. Colourful *you are here!* maps are drilled into walls. Everything is described, pointed out. You can't possibly get lost, at least in theory, because the purpose of every department is made clear to you and on each door hangs an explanation.

The door I was looking for, however, did not possess an explanation. It sat by a reception desk in a small corner of the hospital. It did not have a sign or a purpose drilled into the wall, and should you happen to have noticed it as you walked by, you might have imagined that it led to a stationery cupboard or a small cloakroom. It was the door to the mortuary.

For my time at medical school, I had chosen to commute each day, and during that morning's long drive of one hour and forty-five minutes I had played the radio on an especially high volume. It was a distraction, filling my car and my head with the lyrics to songs. I didn't know it then, but a few hours later, I would return home without the radio on at all. Music would eventually become a barometer, used as a diversion or as a comfort. Its absence denoted an especially difficult day, when I needed the quietness in order to process my own thoughts, and there would be many times when I would drive the entire journey home in absolute silence.

Once I had found the unmarked door, I walked through and was immediately faced with a labyrinth of corridors. In offices on either side, secretaries typed behind towers of patient notes. People passed by me with mugs of coffee and quiet conversation. Everything felt so ordinary. I walked down more corridors, through more unmarked doors, each progressively harder to negotiate, like a real-life but sobering computer game. Swipe cards. Keypads. Codes I had written on pieces of paper were carefully pressed into silver buttons. I became stuck between two doors and had to be rescued by a man in scrubs, but eventually, I arrived at my destination – the mortuary, where I was met by my registrar.

'You made it,' she said.

I think she was as surprised as I was.

The changing room in the mortuary felt reassuring, with its quaint and nostalgic air of a childhood trip to the swimming baths. There were wooden benches and a tiled floor. Rows of lockers, most unlocked and gaping open, were filled with photos and stickers, personalised mugs and spare cardigans. Post-it notes of lives lived outside the hospital. Just like at the swimming pool, there was even a little walk-through area, but instead of being filled with a chlorine solution, pressure hoses and scrubbing brushes hung on its walls. I was given protective clothing, which I expected, but added to that were goggles and wellington boots, and giant rubber gloves which reached right up to my elbows. I stared down at my new outfit and wondered what might lie ahead that made it a necessity.

My registrar turned to me. 'We're now going through those doors.' She pointed. 'Beyond those doors are three tables and on those tables are three bodies, all at different stages of a post-mortem. The body we're working on today is on the far right.'

I stared over at the doors and I felt the familiar tread of anxiety creep from my stomach and into my throat.

'It's perfectly normal to respond to it. It's perfectly normal to feel anxious or upset, or to want to leave,' she said. 'You can leave at any time. Just walk back through those doors and you'll be in the changing room again. No one will mind. No one will think any less of you.'

And with those words, I knew I would be okay. Because with those words, the registrar handed me something very rarely given to us in medicine. She gave me permission to react. Permission to experience emotion and distress, and to acknowledge my own feelings. On so many occasions we are expected to remain impassive, to be mechanised and empty, programmed and preset to be unresponsive to all the unhappiness and misery we encounter. This disapproval of emotional reaction exists in everyday life too. Certain corners of society maintain a particular distaste for anyone displaying emotion, anyone who admits they are overwhelmed or unable to cope. Newspapers and magazines devote a special place on their front pages to any celebrity who cries in public, especially if that celebrity is a man. We are expected to somehow absorb our feelings and our responses to life, to banish them far from the surface of who we seem to be, because their disappearance makes it so much easier for everyone else. In medicine, it's seen as almost mandatory.

'You'll be fine,' said the registrar.

I was fine.

Of course, what I saw beyond those doors was

undoubtedly shocking. My first reaction was that I must be on a film set or backstage at a television show, because the scene in front of me was so far removed from anything I had ever witnessed, and my brain was unable to process it. My eyes needed distance at first, and I prowled at the edges of the room. I wiped my very clean goggles. I adjusted my perfectly placed gloves. I did anything but look at the thing I was there to look at. It's amazing, though, how quickly we are able to adjust and within a few minutes I was at the side of my registrar (who, because she was very wise and very kind, had allowed me to prowl and adjust, and wipe my very clean goggles, without comment).

I'm not sure when it happened, or why, or even how, but at some point during those first few minutes of being in the mortuary, I let go of the shock and the fear. They disappeared somewhere, evaporated into the miracle of what I was witnessing – the miracle of anatomy and physiology, the miracle of the human body, and with each stage of the post-mortem, more small miracles were revealed and more diagrams came to life. When we reached the heart, and the left ventricle was pointed out to me, the sense of excitement was quite extraordinary. There was the left ventricle! The thing I had stared at in a textbook for the past few weeks was right before my eyes! It felt as though I had stumbled upon a celebrity.

The very greatest miracle of all was left until last. As the brain was lifted out and passed to me, I realised that I held in my hands the very essence of who this person was. Their thoughts, hopes, dreams, worries. Their personality. Their sense of self. A lifetime of memories. All of those things rested on my fingers and for a moment, the privilege of what I was doing took my breath away.

When the brain is dissected, you will find, lying deep within the cerebellum (the 'little brain') an area concerned with transporting motor and sensory information. It looks like a series of delicate branches or the fronds of a fern, stretching its tiny fingers out deep within our minds. It helps us to negotiate the landscape, to make sense of what's around us. To survive. It's called the arbor vitae, or the tree of life, and it remains one of the most beautiful things I have ever seen.

After that day, I became a regular at the mortuary. I was on first-name terms with the attendants. I didn't need to check the pieces of paper in my pocket in order to press numbers and letters into a silver keyboard, and I no longer became trapped within corridors. I even had a need for my wellington boots and my elbow-length gloves.

Of course, I still went to the dissection room, which no longer held any fear for me at all, and I would enjoy staring down a microscope at the many towns and villages that lie deep within our bodies. But the mortuary seemed more real, more relevant. Whenever I left, I would move back along the corridors, past the secretaries typing letters and the people carrying mugs of coffee, further and further towards an everyday life, and I would emerge through the little unmarked door at the reception desk. On each occasion, I would walk back to my car, passing crowds of people living their ordinary lives, and I would think 'you have no idea what I have just seen'.

On the way home, I would study everyone around me – a cyclist at a set of traffic lights or people making their way across a pedestrian crossing – and I would imagine all the anatomy that lay beneath their flesh. All those small miracles. I became slightly concerned about myself as I began to wonder if I would ever again view human beings in the same light. I decided I would visit the mortuary one more time. I had other subjects to learn about and it had served its purpose. It had helped me to face the thing that had terrified me the most, and the rest of medical school would be plain sailing.

I was deeply naïve.

On my last visit to a post-mortem, I arrived in the changing room to find my registrar standing in front of the double doors, blocking my way into the tiled room with the three stainless steel tables.

'I was going to text you,' she said.

I thought there'd been a change of plan. Perhaps she had to be somewhere else. Perhaps there were no dead people that day.

'Is it cancelled?' I said.

'No.' She shook her head. 'It's not cancelled. I just wanted to give you the choice.'

I frowned.

'It's a suicide,' she said. 'Do you still want to be here?'

I looked beyond her, towards the double doors, and I wondered what lay behind them.

'I still want to be here,' I said.

It was a man.

The two other tables were empty, and he was alone. He'd failed to turn up at a prearranged appointment and his daughter had gone to her childhood home to find that her father had hanged himself in the garden shed. He was fifty-three. There had been no warning. No prologue to the story. No indication he had decided to take his own life, because you will never spot the suicide who does not want to be seen. The daughter somehow managed to cut him down, and she performed CPR with such desperation, such anguish, that she broke every bone in her father's ribcage. I stared at his face and at the ligature mark around his neck.

We began.

Afterwards, my registrar disappeared briefly to collect something and, for the first time, I was left alone. While we'd been working, someone else had been brought in. It was also a man and he waited for us on one of the other tables. I walked over and looked at the whiteboard on the wall above his head, where the attendants wrote any information they had been given. This man was also fifty-three, and he'd been killed that morning in a car accident. I walked back to my table. Again, I looked across at the other man.

If you had asked me prior to that moment about my views on suicide, I would have told you how much compassion I had towards someone who had been so desperate they had felt compelled to take their own life. I would have told you how much understanding we should all try to find and how we should never judge someone unless we have walked alongside them on their journey. Yet, standing in that mortuary between those two tables, I was filled with so much rage, so much fury, I almost had to leave for fear of being unable to control my temper.

I thought about the daughter, I thought about how she had tried so desperately to revive her father, and about how, no matter what else might happen to her in life, she would never be able to lose her memories of that day. I wondered how the man could have done it, knowing his daughter would be the one to find him, knowing how it would affect her. I thought about how these men were both the same age and had died on the same day, but how one had been given a choice about the matter and the other hadn't.

Many years later, and many patients later, I finally reached the truth.

It took someone else to help me understand – someone I got to know on the psychiatric wards.

He was a junior doctor.

He was also a patient.

Mirrors

My first encounter with suicide was the death of a friend from medical school. Just a letter left to his parents with few words and no answers. No answers but many thoughts. I still knew very little if anything of the effects that the practice of medicine had on doctors. I, like many others, still ask myself if fear of the future was a trigger for his death, the secret thoughts that may have reflected his fears about a career where competition started to shape lives much more directly than a simple mark in an exam did in the past.

The consultant

When Alex was first admitted, he was fractured and confused.

Looking back and reading through the history, as is often the case, you could see the red flags, snapping and fluttering in the breeze, but no one had spotted them at the time. Taking hours to clerk one patient. Sending long, rambling emails to his consultant in the middle of the night. *Increasingly bizarre behaviour*, it said in the notes when he presented himself to A&E, stating he felt unsafe. Self-harm and self-loathing. The wandering, lonely journey of someone who was trying to survive in a landscape that eventually he was unable to tolerate. By the time he reached us, he was showing signs of

29

paranoia – suspicious of everyone around him and refusing to communicate. There were persecutory delusions, perhaps even auditory hallucinations – hearing voices – although we couldn't be certain, because he wouldn't engage with anyone. There were times, though, when he seemed to respond to sounds or to people that no one else could hear or see.

He continued to believe he was working as a doctor for the ward on which he was now a patient. He was very convincing, of course. So much so, some of the other patients began to believe it too.

Slowly, over the weeks, with support and talking, and medication and kindness, Alex began to improve. He started to trust us more. He felt comfortable talking about his thoughts and reactions, and he gained back the insight into why he was there. We had long conversations, Alex and I. He talked about the stress he'd experienced as a junior doctor, how inadequate he felt and how much self-doubt filled his mind every day. I don't think I have ever related more to a patient and their story. I'm not sure that I ever will again. This could have been me. I have always believed that the distance between a doctor and a patient is a short one, but never had it been shorter than with Alex. Most of all, though, we talked about his dog, Fletcher. In another strand that bound us together, he was completely and utterly devoted to his dog. A golden retriever with kind eyes and a goofy walk. He often took out his phone and showed me pictures and videos of Fletcher. While other patients would visit family and friends on a day's leave, Alex would visit his dog in the boarding kennels. Through a tragic twist of fate and the impossibility of geography, he had no family and very few friends. Fletcher was everything to him.

Alex was discharged on a Thursday afternoon in the middle of a heatwave. Before he left, we sat in the shade at one of the benches in the patients' garden and talked for the last time. We talked about different jobs we'd had and the consultants we'd worked for, and we laughed at shared horror stories from the wards. He told me he would like to eventually go back into medicine, because he loved the job and he missed it. He talked about how much he was looking forward to picking Fletcher up from the kennels. It didn't feel as though I was talking to a patient any more, it felt as though I was talking to a colleague.

On the Saturday night, he hanged himself.

I was told at ward round on Monday morning. I had known suicides before, but the shock was so intense, so unbearable, I couldn't find any words at all for a few minutes. When I finally spoke, the first thing I said was: 'No, there must have been a mistake, because he would never, *ever,* have left his dog.'

Because I fell into a trap. I fell into a trap of believing that Alex had a choice, imagining that he sat at home on the Saturday evening and made a decision about whether he wanted to live or die that day, when in reality he had no more of a choice than someone who is killed by a heart attack or by bowel cancer. A disease ended his life, just like other diseases end the lives of people every day. I knew those thoughts were in my head somewhere, but it took me a few days to find them, to realise that choices are not black or white. Choices are coloured and shaded by our own thoughts and experiences, and our decisions are sometimes made – not only by us – but by the diseases that run through our minds.

It was only then I looked back and remembered standing

in a mortuary as a medical student, and being filled with rage and disappointment, and realised, finally, what I should have realised then – that, just like Alex, neither the man who had had the car accident nor the man who took his own life in a garden shed all those years ago had been given any kind of choice.

In reality, what may seem like a choice can in fact be anything but, and it's only when you visit a mental health ward that you begin to realise how small and how rationed those choices might be.

Perhaps, in psychiatry, this is the most important role of all – to restore choice – because the restoration of choice brings with it the return of hope. Many patients arrive on the ward with a complete absence of hope, and a space in their lives where the idea of choice used to live. Choices are born from the acknowledgement of our own emotions, because how can we make a decision about something if we aren't allowed to explore how we feel? The registrar in the mortuary allowed me to explore my feelings about death and by doing so, she handed me the choice of staying or leaving, and the hope that I might be capable of this job after all.

In medicine, and outside of medicine, the need to preserve choice is vital, but perhaps it's felt most keenly of all in psychiatry. A place where choice is so easily lost. A place where, in time, when that choice is found again, there is nothing more rewarding or more wonderful to witness than the restoration of a life worth living. Because the most important ingredient with which to mend a damaged life is hope, for the patient and for the doctor.

Words

Medical students practise suturing on oranges, they practise inserting cannulas into plastic arms and practise CPR on life-size dolls costing tens of thousands of pounds, but there is no way to practise talking to a patient. We bring in experienced actors, we set up imaginary scenarios, and we coach from the sidelines, but nothing can replicate the first time you are asked to deliver bad news. There is no script. There is no one providing encouragement and wisdom from the edges of the room. There is no second chance.

The lecturer

Exposure to the end of life, and acknowledgement of our feelings towards it, is one of the biggest challenges of being both a medical student and a junior doctor. Unfortunately, it is also the topic least spoken about, and dealing with death is a skill we are expected to acquire and improve on with experience, like taking blood or inserting a catheter.

It was one of the first things I noticed when I finally reached the wards. How we pinball from one moment of crisis to the next without time to process our thoughts. How we are expected to move on to the next situation, the next tragedy, without speaking about the one we have just left behind.

How we are expected to carry these parcels of grief around with us each day or learn very quickly how to build walls to shield us from the suffering. But in caring for someone you instinctively begin to care about them, and when something happens to the people you care about, there is no wall strong enough or thick enough to keep you out of harm's way.

For the first eighteen months of medical school, we spent most of our time in a lecture theatre, locked together in the darkness, absorbing anatomy and pharmacology and physiology. Trying to understand the process of disease. Drawing carefully shaded diagrams of the inguinal triangle. Halfway through the second year, however, for one afternoon a week, we were permitted to make the excitable fifteen-minute walk to Leicester Royal Infirmary, where a tired and over-worked consultant valiantly tried to prepare us for life on the wards. We took that fifteen-minute walk many times over the next few years, but we were never quite as enthusiastic as we were on that first journey, our stethoscopes swinging around our necks, a spring in our step. It felt like we were being given a small reward in return for all the hours we had spent buried in our textbooks. For the first time, we felt like *real doctors*, and we said this to each other over and over again as we walked.

The consultant we were assigned to on these precious days in the hospital was a paediatric radiologist. He was wise and experienced. He knew just how to oxygenate our excitement without letting the fire get out of hand. We gathered in a small room off one of the wards and he allowed us to taste dilemmas and scenarios, case studies and food for thought. We could hear patients in the background. Real patients, only a few feet away. We were giddy with excitement.

'Imagine,' he said to us one day, 'that you are seeing a patient about an unrelated condition, and he mentions to you that he has a pre-existing diagnosis of lung cancer. What would you say to him?'

Our nervous little group of eight all side-eyed each other. I was the oldest. I was supposed to make a fool of myself first.

'I'd tell him I was very sorry to hear that,' I said.

The consultant frowned into my very soul. 'No. No, you would not. In fact, that's the very last thing you would say to him.'

I made a small attempt at arguing. As a second-year medical student, I didn't have very much, but I did have buckets full of compassion to throw around to make up for an absence of actual knowledge. Doctors were supposed to be kind, weren't they? Empathetic? What on earth could I say, if I couldn't say I was sorry?

'You would say thank you for telling me that information,' the consultant said. 'Saying you're sorry is a value statement. Those are heavy words, and you might be giving him a weight he is unable to carry.'

He was right. Of course, I know now, he was right. Back then, though, I couldn't understand why saying you're sorry about something was such a problem. Now I understand. Now I understand that each word we give to someone else carries its own burden, and one person's light-as-air is another person's unbearable cargo. We each measure words with different scales.

I learned a lot about the measuring of words as I travelled through medicine. As a (very) junior doctor, I once had the difficult task of telling a young man (and his family) that he had a diagnosis of schizophrenia. It was a diagnosis made by someone much wiser and more experienced than I was, but thanks to a combination of flooded roads and a prearranged, urgent meeting, I was the one bearing the news. I did the best I could. I remember that they were all, understandably, very upset, and I remember saying to this young man that he was exactly the same person as he had been five minutes ago. Nothing had changed, I had just given him a word. He was still the person he was before. But of course, he also wasn't. Because with that one word, I had given him more weight than anyone should ever have to carry in a lifetime. Because words are never, ever, just words.

A few years after my conversation about the dangers of saying you're sorry, I was sent to a county hospital on a final-year medical school rotation called Cancer Care. It was a five-hour round trip, which always gave me plenty of time each day to reflect on 'cancer care' and how the meaning of those words isn't quite as obvious as you might think.

As a student, one of the challenges of being at that hospital – at any hospital – was Finding A Patient To Talk To. It's what medical students do for most of their time. They circle the wards desperately looking for a patient who is willing to tell them a story. It's a way to practise history taking, to understand investigations and medications, and treatment plans. When you arrive on the wards, role play becomes reality, and the page in a textbook becomes someone's life. Talking to patients is the very best way to learn, but it isn't always easy.

On one particular day, I had just about exhausted the entire oncology department (and the outpatient clinics and the chemotherapy suites) looking for A Patient To Talk To and, in a last-ditch attempt, I approached the friendliest-looking nurse on the ward and asked if she knew of anyone who might put up with me for ten minutes.

The nurse glanced around, shook her head and said she didn't think so.

'What about the woman in the corner bed?' I said. 'The one who's knitting? She looks like a possibility?'

The nurse stared at me for a moment, and then she reached into the trolley and handed me a set of notes.

The woman in the corner bed had end-stage bowel cancer. She had exhausted all the treatment options and was being cared for by the palliative team. With the help of the Macmillan nurses, and the skill and expertise of everyone working in community healthcare, she was being discharged. The woman in the corner bed was going home to die.

After I'd finished reading, I looked up at the nurse. 'It would be selfish to ask, wouldn't it? To waste her time?'

'It's not that. She'd be only too happy to speak to you.'

'Then …?'

'You can talk to her,' the nurse said. 'But only if you promise not to say the word "cancer".'

'Pardon?'

'Or malignancy or palliative or tumour, or even growth. None of those words. She doesn't want to hear them. She refuses to hear them.'

'Then what words do I use?'

'All the other words,' said the nurse. 'All the tens of thousands of other words in the English language – just not those ones.'

The woman in the corner bed was indeed very happy to

talk to me. Her husband, however, stayed silent. He had obviously arrived at her bedside straight from work, the remains of his day lying in the folds of his jeans and carved into the leather of his boots. He watched from the chair next to her. The woman chatted non-stop, although the knitting continued as we spoke. There was knitting all over the bed. Wool in every colour you could imagine. Backwards and forwards the needles went, clicking away at all the other thoughts and dismissing them.

We talked about many things. We talked about books and television programmes and holidays. She told me she was knitting baby clothes, because the woman in the corner bed had just found out that her daughter-in-law was pregnant. I swayed for a while on the edge of that topic. It would have been so very easy to fall into it. To colour in future scenarios, scenarios we all knew were not destined to happen. I resisted, because I knew I would be voicing those scenarios, not for the comfort of the woman in the corner bed, but in order to make my own, carefree life just that little bit easier.

'They've been married a year,' said the woman. 'A year last August.'

'Have they?' I said.

'Around the same time, they told me I had a problem with my bowel.'

It hid itself among all the other words, but the story was there. You just had to listen for it.

I could see the woman's husband lean forward just a fraction of an inch.

'But I've had all my treatment now, and I'm going home,' she said. 'Although the Macmillan nurses are going to be helping. Just for a while.'

'Oh, Macmillan nurses are incredible,' I said. 'They were so lovely when my dad was ill.'

I did it.

I fell.

I was doing so well, and yet I stumbled over a piece of my own misplaced kindness.

The knitting needles stopped.

'And how is your dad now?' said the woman.

I hesitated. I looked at the woman's husband. The exhaustion in his eyes. The unforgettable, unmistakeable look of someone who has a complete absence of hope, and I realised that he was the one who gathered up all those unwanted words and carried them around with him each day all by himself.

My dad would have understood. He would have forgiven me.

'He's fine,' I said. 'He's doing really well.'

When I arrived back at the hospital the following Monday, the woman had been discharged. Someone else was in her place – another story, another set of words – and I began to circle the wards again, looking for a patient to talk to. I took many histories as a medical student, but the woman in the corner bed taught me more about the power of words than anyone else. How saying you're sorry isn't always the kindest thing to say. How some words are so heavy that, whether you mean to or not, handing them over to someone else can change a person forever. How – medic or non-medic – we should all choose our words with more care because we never know the scales with which they will be measured.

Medical students are always looking for someone to tell

them a story. If you ever find yourself lying in a hospital bed, you will almost certainly be approached by at least one. Try to accommodate them, if you can. They will be unsure and nervous, and they will falter over their words, but they will be deeply grateful you took the time to speak to them. It's a way to practise history taking, to understand investigations and medications, and treatment plans. When you arrive on the wards, role play becomes reality, and the page in a textbook becomes someone's life.

Talking to patients is the very best way to learn.

But it isn't always easy.

The Wrong Kind
of Kindness

As a doctor, one of the most important things to realise is that patients will always remember how you treated them. Even decades later, they can immediately recall the way a doctor spoke to them or looked at them, and how those words and looks made them feel. I know this because I have been a patient myself, and I too remember very clearly how it made me feel.

I very often think about my car accident, usually when I'm driving. I have no idea why. I very rarely need to travel along the road where it happened and it's not something that is always on my mind (although, like many major life events, I'm sure it's there somewhere, hiding behind all the other thoughts).

Sometimes, I mention it in passing, usually when someone suggests getting a taxi or sharing a lift. I have to bring out my usual 'I was in an accident years ago and I don't like other people driving me anywhere' speech, and I have a familiar surge of anxiety each time I am involved in a decision about how to get from A to B. Sometimes, though, as I drive across the bleak hills of North Derbyshire, with not another car in sight, I start to think about it.

When you go through an incredibly traumatic experience,

something that alters the course of your life forever, your brain seems to sieve out all the big details and it leaves you with just a residue of tiny memories. Sounds. Smells. Textures. I don't remember the accident itself, but I remember (very clearly) the moments beforehand. I remember driving along a straight country road and being overtaken by a number of cars because I am, thank heavens, a very slow driver. I remember climbing a hill. I remember it was a cool, clear evening and I remember wondering what I might have to eat when I finally got home.

The next memory I have is of opening my eyes and realising that the car wasn't moving any more. I was perfectly still. Right in front of me, only an inch away, my headlights lit up a dry stone wall. It was so bright – like the stage in a theatre – and I studied the blanket of moss that covered the surface of the stones. Tiny and flowerless, yet so beautiful. How have I never noticed this before, I thought, on a journey I travel every day. But I knew something was wrong. I knew I shouldn't be facing a different way in the middle of a road, marvelling over the beauty of moss, and so I reached out to put on my hazard lights. That was the moment I first saw the blood on my hands.

I must have lost consciousness again, as I was trying to make sense of everything, because when I next opened my eyes, a man was standing by the car. He told me that he was a police officer. He said it was the second accident he'd stumbled upon when he was off-duty. So, I thought, I have had a car accident. I wanted to ask him questions, all of the questions, but he didn't look at me the whole time he was speaking, and so I stared at the loose button on the sleeve of his jacket instead, and I thought how easily that button might disappear and be lost forever.

Eventually I was lifted out of my vehicle. They sat me in the front of a police car, and I was left alone, with the smell of chewing gum and hoovered upholstery, trying to arrange the jumbled, concussed thoughts that were swimming around in my brain. There was a hiss of a police radio somewhere nearby and, among the clutter of words I couldn't decipher, I heard it say that this was a fatal road traffic collision. Fatal. I didn't realise then that there was another vehicle involved, and my mind charged into a panic, trying to think who might have been travelling with me, who might have died. I went through everyone I knew. Everyone I cared about. I exhausted every possibility, until I finally satisfied myself that I was alone in the car. But, I thought, if I was alone in the car and it was a fatal accident, then the person who died must have been me. I held on to this thought for a very long time. It was probably only moments, but they were the most terrifying, surreal moments of my life, thinking this must be what it feels like to be dead. Being cold and alone in the dark. Listening to the voices of strangers in the distance.

It wasn't until they put me in the ambulance, until they strapped me into a narrow, blanketed space filled with machinery, that I accepted the fact that I was alive. I was still here. I just didn't realise until much, much later how unlikely that was. Just like the off-duty policeman, the paramedic didn't look at me either. He stared at his boots. He stared out of the tiny window in the back of the ambulance. The window was made out of that strange, frosted glass you get in emergency vehicles, and I remember wondering why anyone would stare out of a window when it didn't offer them a view. I tried to talk to him, but I'm not sure the words ever left my head. He certainly didn't answer. I wasn't in any pain at that point, and the only thing I could feel was a wetness around my mouth. It felt as though my nose was running, and I kept trying to wipe it with the back of my hand.

'Don't touch your face,' he told me.

They were the only words he spoke for the entire journey. At that stage in my life, the one reference point I had for a paramedic was Josh in *Casualty*. The paramedic from my accident was no Josh. In my second book, there is a whole scene involving a paramedic. The paramedic in my book is kind and reassuring and thoughtful, because I think, as writers, we sometimes retell the experiences of our lives and turn them into what we hoped they might have been.

When we reached the hospital, I was wheeled through a waiting area filled with staring, and into Resus, where a cluster of people gathered around my trolley. I couldn't see who they were. I could only see their forearms and the navy blue of their sleeves, their scrubbed hands, and the things they passed over my head. I could see the blur of strip lights in the ceiling as I was wheeled down a corridor to be scanned and X-rayed, and all the time, I was hoping someone would just wipe my nose for me. It was all I remember wanting them to do. I always tried to remember that experience whenever I worked in Resus, how your only view is of people's sleeves and arms, and the blinding fluorescent lights about your head. How terrifying it is.

After it was decided I was stable, the cluster of blue uniforms drifted to the periphery, and I was left alone again. It was then she appeared. The junior doctor. She was very young, perhaps only a little older than me, and she leaned over the side of the bed.

'Don't worry,' she said. 'My friend did exactly the same

thing to her face on some rocks, when she was scuba diving in Greece.'

I remember her exact words. I remember the compassion spilling from her eyes.

'It was awful at first, but you'd never know,' she whispered, 'looking at her now.'

It's just a scratch on my face, I wanted to say. It's nothing. They'll probably put some butterfly stitches in and send me home. Why are you talking to me like this? Why are you looking at me with so much concern?

But I didn't say any of these things. I just stared at her. Because her words made me realise, in that small moment of a stranger's reassurance, that I had turned into someone who needed to be pitied.

I understood, when I became a junior doctor myself. I understood that you are constantly surrounded by people who are far wiser, far more experienced than you think you'll ever be. You feel pointless. Redundant. You feel you have nothing to offer a situation, and so you give the only thing you feel confident in giving. You give compassion. You give words. And you over-give those words in order to compensate for your sense of helplessness. The junior doctor in Resus was just trying to be kind. But her kindness terrified me.

Much later, I realised why she had said it. When I had been taken to a ward, and I had persuaded the nurses to let me go to the toilet alone. When I had stood in that toilet, and looked up into the mirror over the sink. When I saw my new face for the first time, and I took a step backwards in shock, because I thought someone else had walked into the room. I discovered, finally, why it felt like my nose was running. The impact of the accident had crushed the entire bonnet of the car, and I'd gone down on to my knees in the footwell. The crash threw me head first into the dashboard and, in the days long before airbags, I had broken the steering wheel with my face. The hard, sharp pieces of plastic had torn into my mouth and nose, and ripped flesh from the bones. So much so that – had you wanted to – you could have lifted my face from my skull, like a mask. The only reason I wasn't in agony was because there were no nerve endings left to tell me I was in pain.

I think about all of these things as I'm driving, but I don't think so much about the injuries, about the months of rehabilitation and the many years it took me to get used to my new face. I think mainly about the junior doctor in Resus. I think about how her misplaced kindness, with all its best intentions, terrified me at a point in my life when I didn't think I could possibly be any more terrified.

I think about the dangers of kindness.

For many months after my accident, I couldn't eat because of the damage to my mouth. (I had to sip banana-flavoured Ensure drinks through a tiny straw and many years later, I remembered its taste each time I had to prescribe it for someone else.) I couldn't speak, either. Or at least, I could speak, but the noise that came out was completely incoherent to everyone else (even though it sounded perfectly understandable to me). So I was forced to write down all the things I wanted to say. Writing down all the things you want to say is a wonderful exercise. It teaches you to be less grumpy. Less snappy. More mindful. As unhappy and frustrated as I was at the time, writing my thoughts down first meant that I didn't let the unhappy, frustrated words go free without a great deal of consideration. I think that, if we were all to choose the words we speak with as much care as we do the words we write, then the world might be a much more bearable place in which to live.

As much as I love the idea of everyone embracing compassion and being lovely to one another, and as much as I like the hashtags and the wonderful ethos of small acts of kindness, you can't sling kindness around like mud and hope it sticks to the right place. Kind words, like all the other words that come out of your mouth (and your keyboard), need care and placement. You really can have the wrong kind of kindness. Kindness isn't a one-size-fits-all. Kindness isn't a bandwagon to be jumped upon, and while kindness might be one of the most powerful, and most empowering, qualities we possess, without consideration, it can be just as debilitating as the most brutal and well planned-out act of cruelty. Because the echo of a kindness really does last forever, for the good or the bad, and you might find that words that you gave with the very best intentions in the world, will be remembered by a stranger many years later, as they drive over the bleak hills of North Derbyshire, with not another car in sight, on the long drive home.

Rose Cottage

On the first day of medical school, we were told that over the next few years we were going to be trained to treat illnesses, to help patients live well, and to also help them to die in comfort. And we were told to never lose sight of the patient. 'Always palpate the patient's abdomen' said our professor (our anatomy lecturer, a surgeon who would retire that same year). This was his lifetime advice – advice that many years later would go on to be described as patient-centred care. 'Always palpate the abdomen.' The same surgeon that told us to be humble enough to accept that some of the things we were going to learn may be outdated even by the time we finished medical school, to never stop learning and to accept that what we once held as the truth may change. And to say sorry. Those words have come back to me many times over the years.

The consultant

In the third year of medical school I was loitering around the nurses' station, trying to look useful, when I overheard the ward sister on the telephone.

She was calling the porters' lodge. Porters are constantly requested throughout the day and night. They are summoned to move trolleys and people, and machinery. They weave and whistle through hospital corridors with requests for X-ray films and blood results, and nervous patients. But this telephone call was different. This request was quiet and unhurried.

'I have a package for Rose Cottage,' the nurse told them.

I didn't understand at first. Was Rose Cottage one of the administration buildings? Was it part of Estates? Was it some building in the distant corner of the hospital grounds, where secretaries typed and filed all the patients away? If so, why were we sending a package there, and why was it spoken about so quietly?

To me at the time, and to those passing by the nurses' station and anyone else who might be listening in, it meant nothing, but to the nurses and to the porters it was a code. It meant that a patient had died.

The package for Rose Cottage is a body for the mortuary.

As a junior doctor, the first job I was ever asked to do was to certify a patient's death. I arrived at the hospital on my very first day, still warm from medical school. I was fresh and unblemished, as yet undamaged by exhaustion and a sense of hopelessness. Still fuelled by a vision of the kind of doctor I wanted to be. My bleep went off within minutes of arriving, and I answered it with the wide-eyed innocence of a child.

'This is Dr Cannon,' I said, still testing out the shape of the words.

'Could you come to the ward and certify a death,' said the voice on the other end of the telephone.

It was a rubbish first job. I could think of many more jobs I would rather have been given, but I presumed it would be fine. My training, after all, had prepared me.

At medical school, we are taught very little about death.

49

We learn many things about the dying process, we read textbooks about the mechanics that lie behind our final breath and the pathology of the diseases that will eventually kill us all, but we speak very little about death itself. There is a space. A space between an illness claiming its victory and the correct way to fill out a death certificate. A space that contains relatives and upset, chaos and reflection. A space that very often contains pieces of our own self-doubt.

Up until that first day as a junior doctor, I had never met death outside of my own family, other than in the detached, leathered cadavers of the dissection room and in the neat rituals of a post-mortem. As a medic, I had never found myself face-to-face with the end of someone's life, at least not one that didn't rest quietly upon a stainless steel table, but still I went to the ward on that day to fulfil my first task as a junior doctor feeling more than prepared for the experience.

And I did know how to feel for a pulse and how to look for signs of respiratory effort. I did know how to check for the presence of a pacemaker and fill out the death certificate. I had been taught all of this, and I could deal with it.

But what I couldn't deal with, and what I didn't know, was how I would feel walking into a room at the end of someone's life and seeing all the small details around that room that told me who this person was. The small details that told me this person's story. The bag of knitting and the get-well cards, the half-eaten packet of Polo mints and the puzzle books. It was the paperback on the bedside table that stayed with me more than anything else. Closed shut, its bookmark resting for evermore halfway through a story. I took the sight of that paperback and kept it with me. It joined other small details I collected on the wards as I went through my days, not realising that it was the weight of these details that would eventually break me.

When I arrived on the ward to certify the death, I collected a pair of latex gloves from a box on the wall, and the rest of the bay watched as I disappeared behind the curtains that were drawn around the patient's bed.

I didn't know the patient. I had never spoken to her or been involved in her care, I just happened to be the doctor on call that day, and I just happened to be the one who was summoned when she passed away. As I worked, I could hear the sound of the rest of the ward, as it played out just beyond the paper-thin curtain. It seemed uncomfortable, brutal almost, that ninety-two years of life could finish to a soundtrack of meal trolleys and floor cleaners, and the whirr and click of visitor conversation. When I'd finished everything I had to do, I removed my gloves and I paused. I looked around the cubicle for something else, another task, but everything had been done. Still, I waited for a moment. As a doctor, my duties had been fulfilled, but as a human being I felt as though the end of someone's life needed to be observed in some way. It felt impossible to turn on my heels and just go, and throwing my gloves into the nearest bin and getting on with my shift seemed strangely dismissive of the long life that had just ended right in front of me.

When I finally left the cubicle, the eyes of the ward remained upon me. I glanced around as I drew the curtains closed again. Most of the patients had visitors and they were curious, but not distressed. The woman in the next bed, though, was visibly upset. She held a tissue in her hand, but she didn't use it for the tears that fell from her eyes, she just folded it and unfolded it, over and over, as I watched. I sat in the empty chair next to her bed and waited.

After a few minutes, she looked at me.

'I shouted at her,' she said. 'In the night, I shouted at her to shut up. Now she's dead.'

'You had no way of knowing that, though,' I said.

'She was making so much noise,' said the woman. 'Moaning and groaning.'

I reached out for her hand, and the folding and unfolding stopped.

'You wouldn't shout at people, would you, if you knew they were going to die? I'm not sure how I'll ever forgive myself.'

When the porters arrive to remove a body, the curtains around the other cubicles are pulled to and the double doors to the bays are closed. The body is disappeared, out of sight, conjured away and through the unmarked door in the basement of the hospital, where it becomes absorbed into the rituals of the mortuary. It becomes a package for Rose Cottage. When the rest of the ward reappeared, shocked and curious, it felt as if those ninety-two years had never existed in the first place.

We do not speak of death. In an age where we swagger at the thought of our own openness, death remains silent, hidden away behind curtains and codes and acronyms. For us as human beings, death reminds us of an inevitability; as doctors, it highlights the fallacy of a perceived weakness. We spend years learning how to mend people, and we line up our armoury of drugs and drips, and machinery, and we rage and fight and argue with death until the very end, as if it acts as some kind of a barometer of our usefulness. Unlike clinicians of years gone-by, society refuses to speak of the end of life as a natural progression, because to do so would threaten our own identity.

Our stubbornness comes with a price. We have rehearsed conversations with patients about dying:

In the event of a cardiac arrest, would you like us to attempt to resuscitate you?

Yes, they cry, yes! YES! Of course they do. Anything else is unthinkable. The gaps in difficult conversations are filled with new treatments and drug trials, and hope. Death has become an adversary, dying has become a battleground. Patients make treatment choices based on these conversations, and our failure to speak openly and honestly means there is a danger of harming the very people we are trying to help. There are no soap opera deaths. Countless times have I chased around a hospital looking for a consultant to sign a *Do Not Attempt Resuscitation* form, because a patient has deteriorated, and there were no plans put in place. Death can be loud and messy and chaotic, and the nursing staff are, once again, left to pick up the pieces. Department of Health research tells us that 70 per cent of people say they feel comfortable talking about death, yet only a handful have discussed their wishes with family, and while most patients state they would prefer to die at home, due to the medicalisation of dying, very few manage to achieve this.

If we are lucky, we will experience the quiet wisdom of the palliative care team. We will be allowed to die at home, or be given a side room. We will have had honest, open discussions where 'end of life care' is said with a sense of choice and empowerment, not an air of defeat. Until we learn how to have these conversations, until we stop talking in codes and acronyms, there will always be patients whose requests are never heard, and there will always be a package for Rose Cottage.

Ninety-two years of life deserves more than the scraping of plates on a meal trolley behind a paper-thin curtain. It deserves more than a corner of a ward surrounded by strangers. It deserves a choice. It deserves some dignity. As medicine becomes more sophisticated, as drugs and treatments become more skilled at keeping us alive for longer, caring for our emotional health as well as our physical health is even more important. A good innings is so very much more than a number.

Spaces

Medical school helped me to diagnose and treat illnesses I would never get to see outside a textbook but failed to prepare me to deal with death. I was told, not even asked, to inform a family of the death of their husband and father following a heart attack, a man that only a moment earlier had been well, just because I was the most junior doctor in the department. I walked the small corridor and broke the news, unprepared, clumsily, badly, not helping those that cared to be cared for. This was my first death as a doctor.

The consultant

In the third year of medical school, you are taken to one side and taught how to break bad news. This teaching involves important suggestions, such as:

Ensure there is a box of tissues handy

and

Make sure you give the patient an opportunity to speak.

There's even a useful mnemonic, just in case you temporarily forget how to be a human being, because there is a mnemonic for everything in medicine, even death.

SPIKES – the handy six-stage protocol for delivering bad news. *Set-up, perception, invitation, knowledge, empathy, summary*. Six boxes to tick, and we practise our conversations around these boxes, again and again, until we are able to fit everything neatly inside. We have workshops too, where we role-play and feel self-conscious, and try out our mnemonics on each other. Eventually, experienced actors are brought in to test out our ability to remember the protocols, but they do not, unfortunately, test out our ability to look death in the eye. The actors leave spaces in the conversation where we can carefully place our mnemonics, because the actors know all about them. The patients we will meet beyond the shelter of a medical school, however, do not.

Towards the end of your medical training, you are expected to start 'working' shifts; that is, to follow a department, not only during medical school hours, but to witness how it operates around the clock. And so, just before finals, I found myself loitering around A&E at a time when I would normally be at home and tucked up in bed.

On this particular night, it was absolutely pouring with rain. It was the kind of weather where people bluster in heroically, stamping their feet and letting out lots of loud gasps. Other than a man vomiting theatrically in one of the cubicles, the department was fairly peaceful, so I helped myself to a few chocolate Hobnobs and sat down with my workbook.

I'd only been sitting there a few minutes when the telephone rang. Obviously, this isn't unusual in an emergency department, but this was no ordinary telephone. This was the special telephone. The one that rings with an old-fashioned bell. And when that telephone rings, it means that something really bad has happened. In this case, the something-really-bad was an eighty-three-year-old lady called Jessie who was having a heart attack.

When the special telephone rang, magical things started to happen. People appeared from nowhere and put on plastic aprons. They brought out lots of mysterious equipment and started writing everything down. If you ever want an example of good teamwork, the resuscitation room is an excellent place to start. Everything is done with breathtaking efficiency. After a few minutes, everyone had been given a role to play and the performance was ready to begin. They just had to wait for Jessie.

When she arrived, it wasn't with the crash and drama of an episode of *Casualty*. It was quietly and almost apologetically through the ambulance bay at the back of the hospital. One of the nurses was doing chest compressions, but she wasn't riding side-saddle like in the movies, and there was no sign of George Clooney anywhere. They flew past me in one giant chain of human beings and vanished through the swinging doors of the resuscitation room. Then I noticed that behind the paramedics and the bags of saline and the red blankets and the chaos was Jessie's husband. He was old and bewildered and wet from the rain. They asked him if he'd like to sit in the relatives' room, but he was too upset. The kind of upset where you want to pace up and down. Half an hour ago, he'd probably been sitting with Jessie in their sitting room, watching the television and thinking about calling it a night. Now he was in a bright noisy hospital and his wife was lying on a trolley, covered in leads and blankets and surrounded by strangers. I've been that kind of upset before and the last thing you want to do is sit down with a nice cup of tea.

They tried very hard with Jessie, but her eighty-three-year-old body had had enough and wanted to leave. I watched them work on her. I watched the drugs going in and the blood being taken out and I watched as they called time of death. I wondered if I would ever get used to seeing people die. Although it's not the dying part that really upsets me; it's the part afterwards that got to me every time.

I knew that, very soon, they would be taking Jessie's husband into one of the quiet rooms, where they would turn his life upside down with one sentence. He had stopped pacing and, when I left the Resus room, he was in the middle of the corridor, staring at the floor and looking for somewhere to be. As I watched him standing there, the rest of the world walking around him, I noticed something for the first time. Hung over his arm, slightly crumpled and with its belt trailing on the floor, lay Jessie's navy blue raincoat.

As a medical student, I had sat and listened many times as patients were told that their cancer was going to kill them. I have heard many consultants fire a warning shot (*I'm sorry, Mrs Jones, but I have some bad news*) and I have witnessed the breathtaking silence that follows these words, as we all wait for the patient to respond.

There is no silence quite like it.

When the patient finally speaks, it's often to offer up a reason to be positive. They quote their friends and the internet and things they've seen in the newspapers. They tell you stories of people they've read about, or people who are related to people they know, or people someone once told them about. People who have defied the laws of medicine. Lastly, when all of these attempts at cheerfulness have floundered, they will give you their final fragment of optimism. The only one left.

'They're always making amazing new discoveries now though, aren't they?'

And then they will leave a pause. A gap in the conversation. A space between all the words where you are expected to place an offering of hope. Sometimes, though, there is no hope to be offered. Sometimes, there is nothing else to be said. The space remains empty and you listen to the sound it makes, as it swallows up everything else.

Medicine is filled with spaces.

Wards and clinics are built on spaces. Spaces in which to put expectation and possibility, optimism and anticipation. We wait in those spaces for test results, X-ray films, scan reports. We prescribe drugs and, as they swim around within a patient's bloodstream, we hold our breath in the space between administration and response. Waiting rooms are crowded with spaces. In a consultation, a patient's anxiety rests in a space the width of a table, as they search for an answer in a doctor's eyes. On the wards, relatives crowd into a side room, looking for a whisper of something they can hold on to, and, in the space between them, in curtained light, hope waits alongside.

As medical students on the wards, in many of our placements we were encouraged to follow a patient from admission to discharge, to take their history and to get to know them, to follow their diagnosis and treatment, and to write up and present our thoughts and learning when the patient is discharged. There is heavy competition to find a patient and many of the 'best' ones are snapped up early. The ideal candidate is interesting enough to fill a presentation, but not so interesting that the workload is increased by having to read up on rare and unusual conditions in case you are asked rare and unusual questions by the consultant in charge.

In one of my rotations, I trawled the wards, looking for a patient to follow. I read through trolleys full of patient notes and scrolled through referral letters on computer screens. I scoured patient bays. I interrogated nurses. On hospital wards, there is a very high proportion of older people. Some 80 per cent of the patient list consisted of people over the age of seventy: those who had been admitted many weeks ago after a fall or with a chest infection – problems long since resolved – but who could no longer return to the life they had once led. They had turned a page. They couldn't manage stairs or their garden path or their lives any more, and so they waited, in wards and bays all around the hospital, for a different beginning. I loved talking to them because their stories were ones that would eventually become silent and disappear forever, but they weren't suitable for a case presentation. I began to wonder if I would ever stumble upon anyone, and I almost gave up, but on yet another circuit of A&E I finally found my patient, lying on a trolley in Majors. His name was Paul. He was thirty-eight.

Paul had been sent there by his GP after presenting with a short history of weight loss and lack of appetite. Vague abdominal pain that reached into his back. A vague feeling of nausea. Doctors' waiting rooms are often filled with the vague and the indefinite, the out of sorts and the ambiguous. Symptoms that can be the result of many different illnesses, some sinister and some untroubling, some that will disappear all by themselves and some that need urgent attention, and it's the GP who has the unenviable task of sorting out the tigers from the pussycats. It wasn't the vague abdominal pain and the vague nausea that made the GP suspect that this might be a tiger – it wasn't even the weight loss or the tiredness, or the feeling that something 'just wasn't quite right'. It was the jaundice.

Every medical student has a list of things to find on the wards. Checkboxes to tick. Clubbing and cyanosis, atrial fibrillation and ascites. Jaundice is also high on this list, and we stalk the hospital looking for examples, our pens poised over our workbooks, like medical birdwatchers.

Do you think the patient in bed four has jaundice? we whisper to each other.

We all walk past several times in covert manoeuvres before we dare mark a tick, not wanting to be fraudulent. We are unsure. Undecided.

Until we see jaundice for real, for the first time, as I did that day in A&E, and we realise there is no mistaking it. No need to walk past a bed several times. No need for hesitancy or self-doubt. When you see jaundice like that, you know you can't possibly be looking at anything else.

I walked over. I loitered at the foot of the bed. Paul looked up and smiled, and his wife looked up and smiled. I introduced myself. I very quickly explained that I wasn't a doctor, because I had learned early on that if you wear a stethoscope and you aren't twenty-one, it pays to make it clear. People always look at your face, not your name badge, which is why Kate Granger's #hellomynameis campaign, which reminds NHS staff of the importance of introducing yourself to a patient, is so vital. I told them I was desperately searching for a patient to present at the end of my rotation, and would they mind if I talked to them?

They didn't. Although, they said, they probably wouldn't be in hospital for very long.

I pulled up a chair. 'It'll only take ten minutes,' I said.

I was there all afternoon.

We went through the presenting complaint, the signs and the symptoms, relieving factors and aggravating factors, and medications and family health, and all the other side roads you wander along when taking a patient history. As you become more experienced, there will be some roads that you may only need to glance down, knowing exactly the route you need to take, but as a student, you very carefully walk the length of each one, worried about missing something important, determined to ask all the questions until you find the answer you're searching for. I wrote everything down. I made a note of the tests that had been ordered and all the observations that bleeped away at the side of the bed. Two little girls played at my feet.

'I couldn't get a sitter at such short notice,' said Paul's wife.

She told me their names and their ages. They were becoming bored, fractious. I went to one of the children's examination rooms to get them a fresh supply of toys. When I returned, they told me about a holiday they had planned for the following week. They owned a caravan in Cromer. Paul was a taxi driver, so he could take time off when it suited them. My dad used to drive taxis. We talked and talked. I think it was a distraction for them, and creating a source of distraction meant that it was one of those very rare times I felt useful as a medical student. But it meant that I didn't spot it. I didn't see what I had done.

There is another space in medicine.

A space that exists between a patient and a medic, and I had walked straight across it without realising.

It's a space that exists for a reason.

Paul was admitted for more tests, and so he and his wife, and their two little children and the toys and coats and carrier bags were all taken from A&E and wheeled through the hospital to one of the wards. I followed them down the long corridors. I saw the surprise on their faces. I witnessed, on the first of many, many occasions, how the very worst day of your life often starts out very cleverly disguised to appear just like any other.

I thought about them on the drive home. I thought about them while I was eating my supper, walking my dog. I lay in bed in the darkness, staring at the ceiling and thinking about them. As a medical student, I possessed very little medical knowledge, but I switched the light on, pulled one of my textbooks from the shelf, and looked at causes of jaundice, and I picked out the ones that suited best the darkness and the thinking. I had begun to consider them as friends, and, as I swam around in the space that lies between the doctor and the patient, I tried to find something I could hold on to.

Every day, I visited them. After the lectures and the ward rounds, after I'd filled in my workbook for the morning and ticked all the boxes I needed to tick, I'd head to the far side of the hospital and check what was happening. I was supposed to, of course, because this was my case study and I needed to keep up to speed, but I knew it was more than that. Even as I walked the long corridors towards the ward and swiped myself through the door with my student pass. Even as I took my workbook out and pulled a pen from my pocket. I knew I was no longer there because I needed a case to present at the end of my rotation. I was there because I cared.

I had heard the conversations at the nurses' station. I had sat in a medical student silence through the ward rounds. I had seen the scans. They had found a mass. There are many small, significant words in medicine, but mass has to be one of the more sinister. A gathering of cells. Interlopers, serving no purpose, tunnelling their way through the body in secret, as we eat and sleep and enjoy our lives, strangely unaware of their existence, until one day, they wrap or cloak or fold themselves around a part of us that we need, and finally we become conscious of their presence.

Usually, on a scan or an X-ray, these abnormalities have to be pointed out to students, because we are as yet unable to differentiate between what should and shouldn't be present. All the organs look vague and puzzling, and we struggle to relate the images we see on a screen to the textbook life we have been used to. But this scan was obvious. This scan was *barn door*, as the radiologists are fond of saying. The mass had buried itself deep within the pancreas, squeezing and pressing, determined to flourish, and elbowing into anything that might lie in its path, including the bile duct, which had, in turn, led to the jaundice. More tests were required to determine the nature of this mass and what its intentions might be, but everyone seemed pretty sure, even without them.

This was pancreatic cancer.

'But he's thirty-eight,' I said to the consultant.

'This is life,' he said. 'This is medicine.'

'It could be something else though?'

'*Painless jaundice is cancer of the pancreas until proven otherwise.*'

It was in medical textbooks. I had written it down myself in lectures and workshops, but this didn't feel like a medical textbook any more. It felt like a person.

'But you're not 100 per cent certain?' I said.

'We're 90 per cent certain.'

I took my 10 per cent from him and held on to it very tightly.

More tests were done. Paul's wife brought in a paperback she thought I'd like and, to give them both a break, I often looked after the children in the day room. Sometimes, I fetched Paul a newspaper from the hospital shop, sometimes, I chatted to his wife about last night's television, and, slowly but surely, I inched my way across that space. No one stopped me. No one turned me around and warned me, because no one teaches you about this space in medical school. Perhaps it is assumed that we will recognise it for ourselves, perhaps we are supposed to identify its dangers without any guidance, but while doctors are meant to lean back, far away from the abyss, it is a basic human reflex to reach across – to discover a connection, a common ground, to find something of yourself in other people. I reached across because it felt like the most natural thing to do.

I had no idea it would eventually be my downfall.

There was a multi-disciplinary team meeting. An assembly of specialists. Doctors and nurses, hospital staff and community workers, all gathered in the darkness and staring at the black and white pictures of strangers. Images of livers and bowels, gall bladders and stomachs projected on to a giant screen. No patients were present. Hope and possibility were passed back and forth across the room. Predictions forecasted. Statistics quoted. Risk assessed. Battle lines drawn and defeats accepted.

Paul was on the list that day, and I sat in the corner of the room in a hard-back chair, waiting in the dim light with a cotton-wool throat for our turn. Because it was an *our turn* now. It felt like it had been all along.

Finally, my consultant spoke. He pointed to the screen and a little red dot travelled around the pictures. He explained the landmarks, for my benefit, and – like many names in human anatomy – they sounded like far-flung destinations on a magical journey. The superior vena cava, the common bile duct, the triangle of Calot. The tumour's position had made surgery an impossibility, and, as if that weren't enough, the cancer appeared to have spread – there was a scattering of spots on the liver, shadows on the lungs: an army of cancer cells, silently marching their way through Paul's body. They talked about stents and the possibility of chemotherapy, they said the word *palliative*. They talked about weeks not months.

I gripped the sides of the chair. There must be something they could do, all of these experts – all of this intelligence, all of this resourcefulness crowded into a curtained room – and yet no one could suggest anything more. The image on the screen changed to someone else and they moved to the next person on the list.

My consultant sat down and Paul's moment was over. He was thirty-eight. He'd never smoked, he hardly drank. He ran the London Marathon four years ago. He had two little girls and a caravan in Cromer. He liked Monty Python and he played football every Saturday morning with his friends. His wife was called Julie and they met on a dance floor in

Birmingham quite by accident, in the summer of 1996, when beer was one pound and seventy pence a pint.

I held all these thoughts in my head because I knew these were the things patients say when they're told a diagnosis: offering them piece by piece to the teller, as if evidence of the unfairness and the unlikeliness of it all will make the diagnosis realise its mistake, change its mind and walk away. These were not my words to speak, this was not my battle to fight, and so I left. I couldn't go back to the ward because of a strange sense of fraudulence, and so I wandered the hospital for the rest of the day. I sat in corridors and coffee shops. I listened to fragments of conversation, walked past the splinters of other people's lives.

A hospital is like a small town. It has shops and a bank, restaurants and a florist. It has a resident community and a wandering population of those who just journey through, and, right at that moment, I couldn't decide which one I wanted to be. I just knew I had to keep walking, trying to find my way across that great abyss, back to being a doctor. It was only when I'd finished walking, along corridors I had never walked before, past doors and departments I had never seen, that I finally arrived at the truth. Once you have crossed that space between doctor and patient, no matter how hard you might try, you will never be able to navigate your way back home again.

My consultant asked if I wanted to be there when he told them. I said no. There is nothing more distasteful, nothing more selfish, than the appropriation of someone else's grief, and I was worried that once I was in the room, I wouldn't be able to hide my own selfishness.

I watched from the nurses' station as Paul and Julie were shown into a side room, a tiny space off the main corridor to the ward. I had been into that room before. It contained four easy chairs and a coffee table, and it was so small it took a huge amount of concentration just to avoid brushing your knees with your neighbour's or accidentally digging into someone else's ribs with your elbows. I wondered how a room that size was going to be able to hold the huge amount of agony it was about to receive. A Macmillan nurse was the last to go in and my consultant looked straight into my eyes as he closed the door behind them all.

They seemed to be in there forever. I wandered around the ward. I talked to other patients I'd got to know. I ticked a few boxes in my workbook. I decided it was better to disappear into another part of the building and stay completely out of the way, and as I walked back down the corridor towards the main entrance, I glanced at the closed door of the side room.

There are some rooms in a hospital that are designed for delivering bad news or made especially for people to sit in while they wait to receive it. The rooms near Resus in A&E. A corridor of small rooms in ITU. A soft, quiet room on the maternity ward, away from the balloons and the cribs and the congratulations banners. These rooms are used for other things as well, of course. They can be used for explaining and planning. Occasionally, you will catch a junior doctor in one of these rooms, eating their lunch and practising a presentation. Sometimes, they are used for giving good news to a patient, but good news is usually delivered at the bedside. Good news is allowed to wander around freely and stretch its legs. It's allowed to travel through cubicle curtains and make its way around the ward and be heard by anyone who might happen to walk by. It's bad news that needs to be contained. Trapped. Kept tightly enclosed in a small room with four easy chairs and a coffee table, just in case it should manage to escape and be heard forever.

I returned to the ward half an hour later. The door to the side room was still tightly closed.

'They haven't come out yet,' said a passing nurse, because she knew me well.

Half an hour later, Paul and Julie emerged from the room. They were different people, because misery always breaks you, and even though you will eventually manage to fit the pieces back together, you will never look quite the same again. They moved slowly back across the ward, Paul and Julie, the Macmillan nurse and the ward sister, and the curtains were drawn around the bed. My consultant sat at the computer next to me and typed something into the keyboard. He didn't make eye contact. When he'd finished, he stood up and said 'what a lovely family' and he left. I wondered, even with wisdom and experience, even with the many miles of hospital corridors he must have walked, if he sometimes couldn't help but take a few steps across that space himself from time to time.

I didn't speak to Paul or Julie until the next day.

'Did you know?' Julie said.

I shook my head. 'Not the last time I saw you. Not until just before you were told.'

'It doesn't make any sense.'

She spoke the words as a question and searched for an answer in my face. I would see that search many times when I became a doctor, the natural assumption that someone with knowledge must also possess a solution, an explanation. The forgivable belief that, as well as understanding the anatomy and the physiology, we are also given the key with which to unwind something and make a life turn back into what it used to be.

'He was as fit as a fiddle,' she said. 'He ran the marathon. He played football.'

'I know,' I said. 'I know, I know.'

We looked at each other in a shared disbelief and I felt this huge need to apologise. For myself and for my ignorance. For medicine's inability to save her husband. For my own selfish misery, which I am certain spilled into my eyes.

Paul watched us from the bed. He looked more jaundiced, thinner. Less likely to survive, although perhaps it was the absence of hope that I was seeing for the first time.

'We'll have no long faces here,' he said, and I watched him put all his effort into making us both feel better, smoothing down an easy road to walk along, taking each obstacle and carefully placing it to one side.

Terminally ill patients do this frequently. I sometimes think they use more energy helping everyone else to deal with the situation than they do in coming to terms with it themselves.

They tried stenting, to find a way around the blockage, but it failed. They tried chemotherapy, but he couldn't tolerate it, and I was just finishing my placement when the palliative care team organised a bed at the hospice.

I had already met patients who had stayed with me, people who remained in my thoughts long after they had left, but until that moment I had managed to continue walking forwards. I somehow always pulled myself away from a dark corner of thinking and directed my energy into the next patient, the next cubicle. This time, though, the stent failed.

I couldn't find a way around it. This time, I couldn't move on. I knew where I'd gone wrong, I knew I'd walked a path I shouldn't have walked, but equally, I knew I would walk it again. Again and again. When I wrote a card for what would be an elderly couple's last anniversary. When I ran around the hospital hunting down ice, for a dying woman in A&E who craved cold water. When I smuggled fish and chips on to the ward, for an old man who had lost his wife and couldn't face eating. None of these make me special or unusual, thousands of medics and nurses do things like this every day. It's what makes us human and, sometimes, walking across that space towards a patient is the only thing we are able to do for them.

I watched them leave the ward, Paul and Julie, and their two little children and the toys and the coats and the carrier bags. An ordinary life suddenly made unordinary by a cruel disease so quiet and so devious we are often unaware of its existence until it's too late. I knew that all the words you normally attach to a goodbye were worthless, and anything else I said would be for my own comfort and not for theirs, and so we just smiled at each other.

I was planning on telephoning the hospice a couple of weeks later. I even got as far as writing down the number and keeping it in my pocket, but I didn't ever get around to calling. On my next placement, I was walking in a different part of the hospital, towards a different ward, when I spotted their Macmillan nurse, on the opposite side of the corridor. I thought of asking her how it went, I thought of finding out the end of the story, even though I knew what the end of the story would be. She recognised me and she smiled. I didn't ask.

I didn't ask her because sometimes we need to leave a

space. A space between a doctor and a patient. Between one person's story and our own.

A space where we can put all the hope.

Beginnings

I have a box full of paracetamol in my bedside drawer. I am so scared of letting my parents down, I worry I might not be able to stop myself from taking them all.

The medical student

I have driven home many times after a night shift without feeling safe to do so. The rota doesn't allow enough time between working days and working nights for your body clock to adjust, and so you feel permanently tired. It affects my judgement – not only when driving, but when dealing with patients. Occasionally, I have slept for half an hour in the staff car park before setting off, because I'm worried I will kill myself or, worse still, kill someone else. Everyone is the same, but you can't speak out because you would be seen as a troublemaker.

The junior doctor

My experience after medical school was marked by compulsory military service and of meeting many others of very different backgrounds who saw me not just as me but as the doctor. For the majority, a figure of comfort, but for some a figure with power that needed to be controlled or used if not abused in order to get something – control or special privileges. The period that taught me that conflict was part of being a doctor.

The consultant

Medical school was over.

Exams were taken, dissertations were handed in, practical stations were tested. For the past few months we had all lived in a knot of anxiety. Waking in the night to check something in a textbook, breaking under the weight of so much knowledge, yet secretly believing we knew absolutely nothing. Blistering in the heat of so much pressure, not only our internal pressure but the unintentional external pressure from those around us.

Each time I sat an exam, I would ring my mother as soon as I came out.

'How did it go?' she would ask.

'It was awful. Everyone said how awful it was.'

'Yes, but … you do think you've passed … don't you?'

If I had failed and I was the only one who needed to come to terms with it, it wouldn't have been a problem, but – like everyone else at medical school – it was the spectators I was worried about. All the people who had stood at the side of the pitch for five years, cheering us on. The parents and husbands and wives and friends. All the sacrifices and the understanding, the support and the kindness. We wouldn't have failed just an exam, we would also have failed the people we cared about most in the world.

That pressure split apart the fracture lines that had already begun to appear. We tried to hold each other up through the revision and the self-doubt. Looking out for those around us. Trying to spot when someone was falling. We weren't always successful.

If you had asked us on that first day of medical school, if you had questioned us about the journey ahead, we would have told you it was going to be the longest one of our careers. Five years sounded like a lifetime, but it was finished in the blink of an eye. I walked to my car after the final exam, back to the place where I had parked every day for the past five years, and I sat for a while in the silence. I had, hopefully, reached the finishing line – although I wouldn't know if I had passed my exams for a few weeks – which meant I might be driving home as a student for the very last time. It felt like there should be balloons and banners all the way down the A50. Instead, of course, the world continued with its day all around me and I went home to wait for the outcome.

I didn't go to the medical school on results day. Instead, I opted to stay at home and receive an email. Most people went in, however, and slowly a trickle of posts began appearing on social media. Champagne and hugging. Tears. Smiles. Joy. Relief. Finally, my email came through. I hunted down my mother.

'I'm a doctor!' I said. 'I'm a doctor!'

As though it were that simple and, in the space between one second and the next, we all became someone new. There is no other degree that changes your identity, that somehow manages to alter your perception of who you are. There is no other degree that carries with it such a glorious history and that ushers you into its celebrated, but occasionally notorious, fraternity. Alexander Fleming, Joseph Lister, Elizabeth Garrett Anderson, Christiaan Barnard. Conan Doyle and Keats.

There were doctors whose impact was so great that we

borrowed their names and gave them to the conditions we had just learned about. Hans Asperger. Burrill Bernard Crohn. George Huntingdon. Alois Alzheimer. We looked at ourselves in the mirror, brand new and shiny, barely harvested junior doctors, and we wondered if we would make even a tiny fraction of that impact.

The next time I went to the medical school, it was for graduation day.

I had been chosen to read the Declaration of Geneva. I have no idea why they picked me. I certainly wasn't representative of my cohort, because, as well as my age, each evening for the past five years I had got into my car and I had driven away from a large part of medical student life. I didn't live in halls. I'd never set foot in the Student Union. I belonged to no societies or clubs and, although I had enjoyed the time I did spend with other students, our lives weren't knitted together in the same way they might have been.

Reading the Declaration still remains one of the proudest moments of my life. I stood in a hall filled with hundreds of students and families, teachers and lecturers, and I led the doctors' oath, a version of which had been repeated hundreds of times over the decades. We recited the words, but we couldn't understand the true impact of their meaning or the many different interpretations which could be drawn around them.

I solemnly pledge to consecrate my life to the service of humanity

I will practise my profession with conscience and dignity
I will maintain the utmost respect for human life.

It was a grand auditorium, with red velvet seats and a

pipe organ. Giant beams stretched above our heads. Gowns and mortar boards scattered over the seats and spilled out into the hallways. Parents cried. Everyone clapped. As each name was read out, there were cheers and shouts. We repeated our oath and made our pledges with the utmost sincerity. Each sentence was meant, each vow considered, but words are always defined by their landscape, and words said in a grand auditorium with red velvet seats are very different to words remembered in the rush of a crash call or at the bedside of a dying patient. We thought we knew what those words meant, but their meaning would evolve with every step we took as doctors. There would be times ahead when we would need to turn away from our consciences. Over the years, the meaning of dignity and respect would be examined and we would take their definitions and turn them over in our minds. In our darkest hours, we would even begin to question the existence of humanity itself.

Since our graduation, the Declaration of Geneva has also altered and evolved. It now contains the vow: *I will attend to my own health, well-being, and abilities in order to provide care of the highest standard.*

Perhaps of all the vows and promises this is the most difficult one to keep.

For all their training, for all their knowledge and expertise in sustaining good health, attending to their own well-being is something that doctors are not especially good at. The focus is always on the patient, solving the puzzle of the person in front of us, not resting until we find our answer. Sacrifice and the surrender of the self are woven into the job, and going without food and water and sleep are also vows every junior doctor seems to be expected to uphold. *Look after yourself,* we are told and then we are placed in a situation where self-care is impossible, and even seen by some as unpleasantly self-indulgent. *Protected mealtimes* say the leaflets we are given, as the bleeps and the phone calls and the requests never stop. *Get home safely* they advise us, as more than half of junior doctors report accidents or near misses when driving or cycling home from a night shift, purely due to sleep deprivation. In a 2017 survey, a third of doctors reported having no rest facilities at the hospital where they worked, and in one of my jobs, all the beds were removed from on-call rooms in order to prevent doctors from taking a break.

I started hallucinating on the motorway, one junior doctor told me.

I stopped at a set of traffic lights and the next thing I knew, a driver had pulled alongside me and the sound of his horn woke me up, said another.

Since 2013, at least three trainee doctors have died in car accidents following night shifts. At the inquest into one death, it was reported that the doctor was singing on the drive home, to try to keep himself awake.

A few years after graduation, I found myself working in A&E as a junior doctor. I had been there for twelve hours without eating or drinking – something that wasn't unusual and was certainly not considered an issue by anyone in the department. However that day, perhaps through the accumulation of many days filled with twelve-hour shifts, I began to feel faint. Waves of nausea tumbled through me and a rushing sound began to fill my ears. My hands were shaking too much to even write in the notes and I certainly didn't trust myself to take blood from someone or insert a cannula. It was affecting my judgement, my reaction time. Extreme tiredness and lack of food have the same effect as being drunk and I was worried that I would make a mistake.

I finished dealing with my patient and I looked around for a chance to escape, just for a minute. The department was in chaos. Every bay was filled, every bed occupied, and a line of paramedics and their patients snaked down the corridor, through the double doors and into the car park. The sense of guilt and shame was overwhelming, but the sense of imminent collapse overwhelmed me just a fraction more. I could be back within three minutes so I had just managed to talk myself into making a run for it when my consultant appeared and told me which patient I needed to clerk next. It was late at night and I knew the canteen would be closed in ten minutes. I didn't want a three-course meal; I didn't even want a sandwich. A bar of chocolate. A biscuit. Something I could eat while I worked. Something to bring me back to being able to function safely. Something to make me useful again.

'I really need to eat something,' I said in a very small voice.

He stared at me. 'There are patients waiting.'

'It's the patients I'm thinking of,' I said.

The look of disgust on his face was so clear and so obvious I can conjure it up at a moment's notice even now, years later. I never asked for food again.

Medicine is a vocation, not a job, we are often told. The reality is, it is both, but when the conditions of the job become unbearable, when the demands made of us are likely to put our own lives at risk, not to mention the lives of the patients in our care, we are expected to continue to bear it because of a deep-rooted sense of purpose. A calling to serve and heal, and to fix. Or perhaps we are drawn to fixing others, because, by doing so, we might inadvertently succeed in fixing ourselves.

Graduation felt like the end. It felt as though we had arrived at our destination and the journey had reached its conclusion. Little did we know that the past five years had just been a prologue, a short introduction to what lay ahead. In our minds, we had passed the finishing line, but in reality we had only been walking, very slowly, towards the start of the race.

In a beautifully circular pattern, the same professor who had delivered our opening lecture at medical school and who had welcomed us to the first day of our medical career, also delivered to us the final words we would hear as students.

He stood on the stage in the grand auditorium, he leaned on the podium, and it felt as if he managed to look each one of us straight in the eyes.

'Now,' he said, 'the hard work really begins.'

Once again, he was right.

Harvested

There is a running joke that you should avoid being admitted to hospital at the beginning of August, because that's when all the new doctors arrive on the wards. In truth, it's the very best time to be admitted, because what new doctors lack in experience they make up for in enthusiasm and compassion. They are yet to be worn down by frustration and tarnished by a broken system. They answer their bleeps immediately, they have time for everyone. They care. There are a few who think nurses are beneath them, but we have ways of putting them right on that score.

The nurse

On a bright, sunny morning in August, the machinery of the NHS turns and all the junior doctors change jobs.

Amid that change, a new harvest arrives, filled with enthusiasm. They are processed, inducted, initiated. They collect bleeps and lanyards, swipe cards and pagers, and they disappear into the wards and along the corridors, where they are swallowed up into the hospital.

I had spent the previous two weeks 'shadowing' my predecessor, a young woman grey with exhaustion and worn down at the edges, who tried to pass on a stream of insider tips for survival, as a parent would to a child.

'The vending machines upstairs never work,' she said.

'Don't rely on the cashpoint near the porters' lodge, because it's always broken.'

'The nurses on Ward 4 are the nicest. They'll make you a cup of tea.'

She told me which consultants were religiously early, and which consultants started a ward round ten minutes before you were due to end your shift. Which consultants you could go to with a problem, and which were best avoided.

'Don't go near her when she's wearing black,' was all she said about one.

She showed me the telephone system and the layout of the wards, how to order an X-ray and how to check blood results on the computer. Extension numbers, requests for porters, paperwork and pharmacy. What to do if you get a needle stick injury. Where the death certificate book is kept. The best parking places. The quickest route from the mortuary to the doctors' mess. I wrote some of the things down and the rest I tried to commit to memory. Like a Labrador puppy, I trotted behind her all day and watched from the safe shelter of the periphery.

On my first day, of course, she had gone.

My job was in urology. A magical mixture of bladders, testes and ureters. There was everything from kidney stones to testicular cancer, enlarged prostates to difficult catheters, and, alongside them, an endless supply of elderly patients who – for a wide variety of usually mysterious and unexplained reasons – had stopped being able to pee.

Surgery days always begin earlier than medical days, and at seven-thirty I arrived for my first shift as a doctor, allowing myself an extra half an hour to find a free computer and print off my patient list, which would eventually turn out to be the first challenge of any morning. The ward round began at eight sharp and this gave me just a couple of minutes to spare in which to congratulate myself for finding the right ward, and for making it on time and with all the various bits and pieces I needed to have about my person.

Surgical ward rounds are fast. Unlike medical ward rounds, which often pause for reflection and conversation, and sometimes last all day, surgical ward rounds are swift, clean and unfussy. Surgeons belong in theatre, and I often wondered if some of them saw speaking to patients as more of a sideline. A minor distraction before the real work began. The consultant coasted through the ward, occasionally stretching out a hand for a set of notes or an observations chart. We scrambled behind him, trying to keep up, arguing with unruly cubicle curtains – which were no sooner drawn to than he was on his way again – and the notes trolley became a vehicle of chaos. We each had a job on the ward round, and mine was to prescribe the medication, which was all done electronically. I could barely remember my log-in details, let alone my PIN number, and I couldn't hear what the consultant was saying above the noise of the trolleys and the breakfast plates. I didn't dare ask him to repeat it. The laptop teetered on the edge of a tower of notes. People moved it to retrieve those notes and it shut itself down. I managed to restart it and log myself in, but it was borrowed and taken away, and someone else logged me out. I tried and failed to put my details in again. The laptop became very angry and locked me out of the system altogether. I began to shake. The senior house officer (SHO) leaned over and pressed a few keys. He typed in all the prescriptions that were needed in a matter of seconds, and when he finished, he looked at me.

'Don't worry,' he said. 'It gets better.'

At the end of the ward round, the consultant disappeared to theatre and the rest of us were left in the wake of a tornado, with a jobs list to divide up between us. There were notes to write, discharge letters to type, medications to change and blood tests to chase. Everything in medicine is chased. *Chase that,* a consultant will say, tapping at an X-ray report or a bloods request, and you will find that your to-do list is mainly comprised of things you are required to run after. It was a list that stretched across the day and late into the afternoon.

My predecessor was right, though. On Ward 4, the nurses do always make you a cup of tea.

By the end of the first week, the SHO's prediction came to pass. It did get better. I learned my log-in details by heart. I knew to be prepared, and to prescribe 'as needed' pain relief and anti-sickness for anyone returning from surgery. I knew to clip consent forms to the front of notes and to make sure no patient was ever sent to theatre without a cannula in place. I began to feel more relaxed. Bedded in slightly. Perhaps even a little bit useful.

Unfortunately, though, medicine never allows you to relax for very long, because no sooner had I completed a week of day shifts than the timetable changed again. I was about to roam the corridors alone for hours at a time, not only called upon for my own patients, but for any surgical patient in the hospital.

I was about to start my first set of nights.

The hospital at night is a different country. The first thing you notice is the absence of sound. In the day, there is a white noise of trolley wheels and telephones, conversations and footsteps, and on every corridor great waves of people move towards you – nurses, doctors, porters, cleaners. The crash and clatter of laundry carts and meal trays. The reassurance of sound. During the day, there is no corner of the building where you are able to find even a small space of silence.

At night, everything is made out of silence. Not until noise is removed do you appreciate how comforting it is to hear the distant work of other people. Apart from a trickle of evening visitors, heading back towards the car park, I saw no one on my walk to handover. The shutters were down on the little shops and at the florist's, and upturned chairs rested on the coffee-shop tables. A row of secretaries' offices, usually an engine of efficiency, all stood in darkness. Computers sleeping. Mugs rinsed on draining boards. Tea towels folded. I passed A&E, a place where the terms day and night cease to have any meaning, and the bright lights and sound of people talking provided a small consolation. I was not alone. There were other doctors scattered around the hospital, and among them, somewhere, was my registrar. The person I turned to if I needed help or if I felt out of my depth. All I had to do was call his bleep and he would be there.

Evening handover was held in a small teaching room on the top floor. I was the first to arrive and I waited in the hum of a strip light, alongside a broken projector screen and a plastic skeleton. To pass the time, I stared at the list of symptoms written on the whiteboard and tried to work out what had last been taught, but the notes made no sense to me. My inability to recognise them sent me into a small panic. What was I even doing there? I knew nothing. I was a fraud. A charlatan. It felt as though, at any moment, a GMC police officer might march through the door and arrest me on the spot for being an interloper. I took a small collection of talismans out of my pocket for reassurance. My stethoscope, a notebook and pen, a laminated card of out of hours bleeps – pathology, radiology, ECGs and the ever-present portering services – and a long list of extension numbers for every ward in the hospital. I took out my tourniquet, which had cartoon bats printed on it and was clearly designed for paediatrics. When I'd bought it, it had seemed amusing, but in the stark light of that teaching room it felt almost grotesque. Lastly, I had a reference book, a handy guide for junior doctors that we all carried around with us: bullet-point lists of what to do and what to check and what to prescribe in different situations, as if all of life's emergencies could be condensed into a small handbook and stored neatly in your pocket.

The outgoing doctor arrived to give me the on-call bleep. He apologised for the long list of jobs he was about to read out, because there is an unspoken rule in medicine that you don't hand over jobs in handover – a rule which is impossible to follow. His scrubs were creased and tired, and he looked as though he hadn't slept for a week. The bleep sat between us. It went off six times while he was talking, because, as I would very soon learn, the silence of the hospital at night is just a smokescreen, and buried somewhere in that silence there is the never-ending sound of small tragedies.

The only other person present in handover was the site nurse practitioner, one of the most experienced nurses in the hospital, who patrolled the wards during the night assessing the most unwell patients, alerting the staff to any problems, sorting out beds and difficult cannulas, ensuring the hospital ran smoothly, and tidying up after the doctors. I didn't know it then but she would be my guardian angel, her name was Claire and the fact that she was Irish just added to the sense of relief. I would have quite liked Claire to follow me around for the rest of my life, giving me a gentle push to believe in myself and a hug of reassurance when I needed it.

After they left, I sat alone studying the list of jobs I had been given. The bleep went off three more times. I rang all the numbers and added more jobs to my list.

A lot of medicine is learning how to prioritise, to decide which patient needs to be seen first, purely from a quick phone call. Is it the man on Ward 4 whose blood pressure has dropped? Is it the woman on Ward 7 with a temperature? Is it the patient who has just arrived in A&E with a short history of vomiting and abdominal pain? I learned as I walked the corridors, and I slowly began to develop an instinct, a sixth sense that told me which patient I needed to see first. Sometimes, I was wrong, but the more times I got it right, the more my confidence began to grow.

I didn't hear from my registrar. I caught a glimpse of him occasionally, when I passed through A&E, but our paths never crossed. As well as developing an instinct about the patients who needed me most, I also learned that there are different kinds of registrars – those who enjoy bleeping you every hour to find out exactly what you're up to, and those who like to walk the corridors alone.

As the night continued, staying awake began to be an issue. While adrenaline had kept me going for the first few hours, a broken, anxious sleep the day before meant that by 3 a.m.

I had started to flag. I foraged from vending machines. Rich chocolate and paper-cup coffees. Crisps and little pots of cream cheese. Even in the brief moments when I hadn't any jobs to do, I walked the corridors to stop myself from falling asleep, and, in an act of desperation, I stepped outside the doors of A&E and took in a huge lungful of cool night air to wake myself up. There were no on-call rooms, no beds put aside for night doctors, and all over the hospital, bodies of sleeping medics littered waiting-room sofas and office floors, as they tried to catch a twenty minute nap, their bleeps pressed into their faces. I didn't dare. What if I couldn't wake up? What if I slept through until morning and was discovered by a cleaner stretched out on the carpet? What if someone died?

The night walked on. I reinserted cannulas in darkened bays. I prescribed sleeping tablets, reviewed patients, took bloods ready for the eyes of the morning teams. I ordered ECGs and put catheters in. I walked cultures down to pathology. Somewhere along those corridors, somewhere between the hours of midnight and six in the morning, I began to feel useful. I felt as if I had a purpose. For the first time, it seemed as though all that training had grown into something worthwhile and I began to feel like a doctor.

There is a point on a night shift when one day turns into the next. It seems to rest on only a fraction of time. The first laundry truck appears. You catch a glimpse of a cleaner along one of the long corridors. The sounds and smells of breakfast drift along the walkway leading to the canteen. It's not just the people and the increase in activity, though, it's more to do with a change in the air, a sense that the building is stirring and waking to a new day. As the giant clock in A&E clicks from one second and into the next, there is a brief moment of limbo – a grey space of nothing – and then morning appears to take over the reins.

I had survived my first night shift. Despite the tiredness, I was elated.

My second night shift went just as smoothly. There were a few admissions, a few unwell patients on the wards who needed reviewing and monitoring, but, other than that, I continued with my on-call staples of cannulas and bloods, sleeping pills and pain relief.

I met my registrar, quite accidentally, in the corridor leading to the orthopaedic ward.

'Everything okay?' he asked.

'Everything's okay,' I answered.

It was the only time I spoke to him all night.

By the third shift, I had shaken off almost all of my anxiety. I parked my car and practically marched towards the hospital entrance. I was smiling, at ease. I was almost looking forward to it.

I didn't realise I was about to experience the very worst night of my life.

The Darkest Hour

I have seen junior doctors bullied by consultants. I have seem them intimidated and persecuted, humiliated on ward rounds in front of patients and nurses, deliberately made to look stupid. Some consultants give nothing but kindness and support, while others seem to go out of their way to make their junior's life a misery. Student nurses are more protected on the wards and they're much safer because they have so many different colleagues they can talk to. Junior doctors are often on their own. We try to take them under our wing, but there is only so much you can do. I often think about the ones I've known – the good ones – and hope wherever they are, they feel valued. Everyone should feel valued.

The ward sister

The night shift began just as the others had done.

A handover in the little room with the plastic skeleton. The same doctor with the tired eyes. Claire, the Irish nurse. The exchange of bleeps and the passing on of jobs. My registrar never came to handover, but he was out there somewhere should I need him, moving around the hospital – dealing with the emergencies and making the difficult decisions, while I got on with the routine and the mundane.

There were only a few jobs given to me, and none of them was urgent, so I sat for a few minutes and looked through a list at the patients I wanted to revisit, people I'd seen on previous nights and wanted to check on. A man with a grumbling high temperature on the male surgical ward. A woman on the second floor with a urine infection we couldn't seem to shift. A dying woman in a side room on Ward 11, who had managed to hold on to another day and still remained on the list. I put the piece of paper in my pocket and made a start.

It was just coming up to 2 a.m.

Everything was going well. I'd just reinserted a cannula, for the third time, in a patient who seemed intent on pulling out all my glorious efforts the minute my back was turned, and I was going to make a pit stop at one of the vending machines when my bleep went off.

It was A&E. It was most likely a new admission, or someone who needed their medication writing up, and I stopped at a phone in the corridor in order to call them. It was my registrar. This was the first time in three nights he'd got in touch with me.

'Could you come down to the emergency department?' he said.

I couldn't imagine what he wanted. Some registrars liked you to write in the notes as they spoke, or prescribe all the patient's medication, but this registrar seemed happy to do all of those things by himself. Perhaps it was a big emergency, I thought. Perhaps it's something important and he needs my help. I quickened my pace, fully expecting to walk into the scene of a major incident as soon as I pushed open the double doors to A&E. There was nothing. If anything, the department seemed quite peaceful. A couple of staff were restocking one of the trolleys and Claire was sorting out a medical bed for a new admission. She turned and smiled at me.

'Everything all right?' she said.

'Yes,' I said. 'Yes, everything is fine.'

I looked around the department and spotted my registrar sitting at the main desk, leaning back with his hands laced behind his head. He gestured to another chair and I sat down.

'Everything all right?' he said.

'Everything is fine.'

'Are the wards okay?'

'The wards are fine,' I said and I frowned at him.

'I'm going to Amsterdam,' he said.

I frowned a little harder. He'd bleeped me to come all the way to A&E purely to talk about a holiday he had planned?

'Right,' I said slowly. 'That's nice, when are you going?'

He leaned forward and smiled. 'Now,' he said.

I waited for the punchline. There was none.

I stared at him. 'What do you mean, *now*?'

He pushed something into my hands and I took it without looking down. 'I mean I'm leaving now,' he said. 'You're in charge.'

There was a wave of anxiety so terrifying it pushed bile into my mouth.

'You can't just leave,' I said. 'I can't be here all by myself!'

He stood up. 'You'll be fine.'

'But there's six hours left of the shift!'

'If I don't go now, I'll miss my flight.'

He started to walk away.

'You can't just leave!' I said again, only this time I shouted it.

Claire, who had heard the whole conversation, shouted to him as well.

He carried on walking. He even waved at us, without turning, like Liza Minnelli in *Cabaret*.

He was gone.

I looked down to see what he'd given me.

It was a bleep. His bleep. The same bleep I was supposed to call if I was in trouble, or felt out of my depth, or needed help. The bleep everyone else in the hospital called if there was a surgical emergency.

I held it in my hands.

It belonged to me now.

I had been a doctor for ten days.

Within three minutes, the paramedics had arrived with an emergency, sirens and lights and a crash of doors. My registrar must have driven right past them on his way out of the hospital car park.

It was a young man with severe abdominal pain and vomiting. He also had learning difficulties. He had a long and tricky cardiac history, and multiple other health problems running alongside, and if that weren't enough to make things complicated for him, and also for us, he had a permanent tracheostomy. He was in severe pain. He was understandably frightened and he was thrashing around on the trolley, kicking and punching each time anyone went near him.

The A&E consultant shouted from somewhere in the middle of a crowd of people.

'Where's the surgical team?'

I felt the weight of the bleep in the pocket of my scrubs and I took a deep breath and swallowed back the bile.

'I am the surgical team,' I said.

I watched from the corner of the Resus room as the A&E staff stabilised him. They monitored his heart and his breathing, they got his pain under control, and they managed to calm him enough to examine him and take bloods. I watched in awe at their skill and their expertise, at their kindness and understanding, but also with a huge weight of guilt and deep anger that this young man was not getting the surgical doctor he was entitled to. He was never put in any danger and he was never neglected, but he deserved better than me. At that moment it felt like every patient in the hospital deserved better than me.

The A&E consultant walked over to where I was standing.

'He needs an ITU bed,' she said and looked at my badge. 'Where is your registrar?'

'Amsterdam,' I answered, because there wasn't really anything else I could offer.

'I'll get him an ITU bed,' said a voice from across the room.

It was Claire and she did exactly that.

I watched the young man being taken away by the porters. He also had a nurse and an emergency department doctor with him, and his trolley was littered with so many wires and so much equipment that it was difficult to tell if there was a patient lying among it all.

The sense of relief made my legs unsteady. He belonged to someone else now – someone far more capable than I was – and I could go back to my own job, because, while the registrar's bleep had stayed eerily silent while I was in Resus, my own bleep had gone off so many times that some of the numbers couldn't be stored and they were erased forever. It wasn't a problem. There was no doubt in the world that they would bleep me again.

I worked my way through the phone calls, prioritising jobs, talking to nursing staff. All the time I thought about the young man and I wondered how he was doing. Ten minutes later, my bleep went off yet again. It was ITU.

'This is the on-call doctor from ITU. We have your patient,' said a woman's voice.

She put a particular emphasis on 'your'.

I hesitated. 'Yes,' I said.

'I just wondered,' said the doctor, 'would you like me to prescribe all his regular medication?'

I hesitated again. Would he be having surgery tomorrow? Was any of his medication inadvisable if he was? What about the drugs he had in A&E – did any of them mean he shouldn't have his regular medication alongside?

I didn't know. All of it would be written in his notes, which were now in the hands of the doctor who was calling me. This doctor also happened to be an ST5, and, in the strangely labelled hierarchy of doctor training, that meant she was at least five years more experienced than me.

'I don't know,' I said.

She put the phone down.

Ten minutes later, she bleeped me again.

'Would you like me to put a nasogastric tube down?' she said. 'To feed him?'

There was silence. It felt like an exam.

'Does he need a chest X-ray?' she said.

Still silence.

'He has a tracheostomy, is a chest X-ray still necessary?'

'Yes,' I said, but I turned it into a question.

She put the phone down again.

This went on through the night. Every twenty minutes I would be bleeped with a question or to tell me a pulse rate or a blood pressure reading; to ask me what she should do, even though her experience meant that she was more aware than I was of what was needed. When she had drawn out my ignorance and highlighted my complete lack of knowledge, she would replace the receiver without speaking. Clearly, this woman was having just as bad a night as I was, but it felt like I was being punished. It felt like bullying.

Twenty minutes later, my bleep went off again. I presumed it was my next set of impossible questions, my next round of punishment, but it wasn't. When I checked the number, it was one of the wards.

'Can you come to Ward II?' said the nurse. 'Can you come straight away?'

It was the woman in the side room, the one on my list who had held on to life for another day. At five o'clock on an August morning, her body had decided it was time to leave.

It was not an easy journey. The cancer she had endured had marched through her body, taking her organs one by one, burrowing deep into her bones and spreading itself throughout her brain. She had pain relief and anti-nausea medication prescribed, drugs to help with her swallowing and her anxiety, but it wasn't enough. I could hear her cries as I walked down the corridor towards the ward.

'Could you give her some more morphine?' said the nurse.

I looked at the woman's drug chart. She was almost up to the limit of what I was allowed to prescribe, but I could risk a little more.

We waited in silence, the nurses and I, sitting in pools of light made by the night lamps on the desk. The crying continued.

Let me die the woman shouted from the side room, *please just let me die.*

We waited. Perhaps it needed time to kick in.

Please just let me die.

'Can't you prescribe any more?' the nurse said.

I solemnly pledge to consecrate my life to the service of humanity.

I looked at the drug chart again. The woman was on the maximum dose allowed. If I wrote up any more, not only would it be illegal, it would look as though I had deliberately put an end to her life. It would look as though I'd killed her. Her relatives were on their way, what would they think if I prescribed too much? What would they think if I didn't?

'I can't,' I said. 'I'm not allowed to.'

Please just let me die.

'Where's your registrar?' she said.

'He's gone to Amsterdam. He's disappeared in the middle of the shift and left me. I don't have anyone else. I'm on my own.'

'Then it's down to you – she needs more pain relief.'

Please just let me die.

I stared at the drug chart. What should I do – should I stick to the rules or should I write up the morphine and face the consequences later? If this was my mother, wouldn't I give her all the morphine in the world, just to put an end to her misery? Was I putting myself and my own survival before the needs of a patient?

I will practise my profession with conscience and dignity.

We rang Claire and a few minutes later she appeared on the ward. She was authorised to prescribe a little more.

'This is all we can give,' she said.

It didn't do a thing. The crying continued. It was like nothing I had heard before – brutal and desperate, and drawn from a place I could never begin to understand. The final cries of someone who needed a release that no one, even with their training and knowledge and ability, was prepared to give to her. This was something they don't teach you at medical school, something you only under-stand when you have walked through it; something you can never un-hear, because I knew, as I sat there in a pool of light at the nurses' station, I would remember that voice for the rest of my life.

'I can't stand this any more.' The nurse got up and walked away.

I will maintain the utmost respect for human life.

I made myself stay. I made myself sit in the chair closest to the side room and I forced myself to listen, because I knew I would always need to remember the sound of my own inad-equacy. Woven into a doctor's sense of self is the need to revisit our own failures, to return to a memory again and again, so that it never loses its colour and its brightness, so we will forever have to hand a reminder of our own flaws and incompetence. Perhaps it tethers us to the limitations of medical knowledge, and our own skills, and stops us from floating away on a fantasy of the person we would like to believe we are. Perhaps it does none of those things and we just feel comforted by the reminder that we are only human. Perhaps, in the end, it makes us better doctors.

It was 5.30 a.m and it would be several hours before anyone more senior arrived at the hospital.

'We need to ring the on-call consultant,' said Claire, and she passed me the telephone.

The voice on the other end was sleepy and faint, but very clear.

'Prescribe as much morphine as you need to make her comfortable,' she said. 'But write in the notes that we've had this conversation and the time that we had it, and write very clearly that I've instructed you to do so.'

Within minutes, the crying had stopped. I listened into the silence, but there was nothing. The woman slept peacefully, her breathing slow and steady. I looked at her face and I wondered where she had gone to.

One of the younger nurses on the ward sat next to me and watched, pale and tear-stained, as I wrote in the notes.

'I will never forget this night,' she said.

I looked up from the page. 'Nor me,' I said. 'Nor me.'

I was just leaving when my bleep went off again. It was the doctor from ITU.

She read out a list of observations, needle-sharp, on my patient. Blood pressure, pulse, respiration rate, urine output.

I listened to her in silence.

'What would you like me to do?' she said.

I felt a tired rage grow in the well of my stomach. It made its way through my body, flowing through my legs and my arms, filling my head and my eyes and stretching right to my fingertips. I gripped the telephone so tightly, I was worried it would break.

'My registrar walked away from the shift in the middle of the night. I am carrying a bleep I should not be carrying. I've just spent the last hour listening to the cries of a dying woman and there was nothing I could do to help her. I have been a doctor for ten days, so what I would like you to do,' I said, 'is to leave me alone. What I would like you to do is to stop bleeping me.'

I put the telephone down. She didn't call again.

I took a journey through the hospital. I chose the longest route, because there are times when your unhappiness and self-loathing is so great there is nothing you can do except try to walk as far away from it as you can.

While I had been on the ward, time had crossed over the second when night becomes day. I could hear the whir of floor cleaners somewhere along a corridor and the rattle of breakfast plates in a kitchen. Porters moved between wards, nurses crowded into offices for handovers, and the next shifts began. The hospital had turned around me and everything else had become clean and new again, leaving me to walk the corridors wearing yesterday's thoughts. It was as if the previous night had never happened, as if the whole thing had been imagined – constructed from my own worst fears, crouched hiding in the corners of my mind.

At eight o'clock my shift ended, and I waited by the staff entrance to the hospital. I watched everyone arrive for work. Doctors mainly, a great sea of them, pouring through the doors and into the corridors.

'Where were you,' I thought, 'a few hours ago? Where were you when I needed just one of you to be here?'

Eventually, I spotted him. My consultant. Smart and

suited. Overcoat and briefcase. I stood in his path and he slowed his pace until he was standing right in front of me. He stared.

I handed him the bleep.

'What's this?' he said.

'This is the registrar's bleep and I have been carrying it for the past six hours.' I could hear my voice falter and break. 'Your registrar left me in the middle of the shift. He went to Amsterdam. I have had no registrar support or supervision since two o'clock this morning.'

I expected outrage.

I expected I'd have to stay beyond my shift and make an official statement, to put something in writing. I expected repercussions, perhaps even an inquiry. At the very least, I expected a reaction.

There was nothing.

'Did anything …' he paused, '… untoward happen?'

I hesitated. What would he define as untoward? No one had died. Everyone had received the care they needed, eventually. But I thought about the young man in ITU. The woman in the side room on Ward ɪɪ. They deserved better, they deserved a doctor who didn't vanish in the middle of a shift.

He took my hesitation as a no.

'Then I don't know what you're bloody complaining about,' he hissed, and he put the bleep in his coat pocket and he walked away.

I stared after him. One of the nurses from the ward was standing next to me. She was leaving after the shift, a raincoat over her arm, a bag on her shoulder. She leaned forward and whispered in my ear.

'He already knew,' she said.

I watched the consultant walk further down the corridor and, in that moment, I realised I was on my own. I was no longer in the safe embrace of the medical school; I was in a job now and this was not a job in which you spoke out. This was a job where the rules were defined by the players and I was clearly expected to keep my head down and my mouth shut. The lines had been drawn around me and, whether I chose to cross them or not, those lines would determine whether I sank or swam.

I knew all of those things because, as I watched the consultant walk away and I saw the arrogance in his step, the way he laughed and waved to one of his colleagues across the corridor, the way he didn't even give me a backward glance, I knew that the nurse was almost certainly right.

He had known all along.

Roles

The hospital ebbed and flowed and I allowed the undercurrent to take me. After that night shift, it seemed easier to be carried along by jobs and bleeps and telephones each day, rather than be prisoner of my own thoughts. The ward rounds and the rotas, the pressure and pull. I tried to stay at the edges, to do what was expected without allowing it to take ownership of me, but eventually this proved to be impossible.

All the way through medical school, the one thing that keeps you going, despite the exams and the travelling, the lack of money and the complete absence of free time, is the idea of what kind of doctor you are going to be. You don't fantasise about prizes and awards and accolades, you imagine the small and the ordinary instead. Having time for your patients, being able to explain a treatment to someone in a way they can understand, helping someone's journey to be a little more bearable. It's only when you arrive on the wards, when you are spat out into an NHS that bends and breaks under the strain of the endless demands placed upon it, it's only then that you realise you will never be able to be the doctor you want to become.

The system simply won't allow it.

Instead, you will carry three bleeps because no doctors applied for the jobs next to yours on the rota. Instead, you will trip over your own misery, as you attempt to keep up with everything you are asked to do. You will watch patients drift past each day who are clearly unsure and afraid, but there will be nothing you can do about it. Relatives wait for reassurance, but go home empty-handed. Waiting lists are full. Clinics are crowded. Everyone pushes and fights and shouts to be heard above the noise of other people's agony. Rights become privileges. Equality becomes discrimination. Time, money, resources and hope all run dry. The NHS is held together by the goodwill of those who work within it, but even then it will fracture, and you will fall into the gaps those fractures create, and you will disappear.

You quickly find that you can never be the doctor you wanted to become, because the doctor you wanted to become would not be able to survive.

As you move with the current that ebbs and flows, you occasionally find yourself washed up in a place where there's a chance that you might make a small difference, and when this happens you fight very hard to hold on to that chance. Perhaps it helps to balance out all the times when the system forces you to walk past, when you have to turn away because the disempowerment you feel is too much to stomach. You reach out with both hands for such opportunities, even if it means that, by doing so, you have to let go of another tiny piece of yourself.

My chance to make a small difference was called Joan. She was seventy-nine. She had learning difficulties and was blind, and she wore a very powerful hearing aid, which Joan was very selective about using, depending on the topic of conversation and whether it suited her or not. As if life hadn't thrown enough challenges at Joan, it had recently been discovered that she had inoperable cancer. It wasn't something Joan wanted to know about, and each time it was mentioned, she would very deliberately turn off her hearing aid and disappear away into another corner of her mind. Joan's younger sister – who was a sprightly seventy-four – was left to make all the decisions and to deal with Joan's infamous temper and tidy up all the pieces of her life, and I think this was probably the way it had been ever since they were children.

Joan was in a side room on a ward at the very top of the hospital. All patients are equal, but some are more equal than others and I fell for Joan immediately: for her feistiness and her independence, her refusal to let all of these challenges get the better of her. I fell hook, line and sinker. I thought she was wonderful.

After my shift had ended, I would call to see Joan. She had refused any palliative treatment – no chemotherapy, no radiotherapy – and she waited, as many older patients do, in a grey limbo for somewhere else to go. Joan needed a nursing home, but her needs were great and very specific, and all the nursing homes we tried didn't need a Joan. Each afternoon, I would stand by the door of the side room and call my name, and Joan would let me know if I was allowed to come in. As time went on, I was never refused.

Joan had never learned Braille, so I read to her from books and magazines. I described the world from the windows of her side room. I brought her chocolate from the hospital shop and I was told off for buying the wrong kind. I made tea (too sweet) and coffee (too bitter). I was always either too late or too early, but my company was never turned down. I got to know Joan's sister, and whenever the two of us talked, Joan would take great pleasure in disagreeing very loudly with anything either of us said. I adored her.

One day, I was on the ward for something else and I stood outside Joan's room and called out my name. There was no answer. There was always a 'come in', or very occasionally an invitation to clear off, but there was never no answer. I peered around the door. Joan sat next to the bed. The radio, her usual companion, was silent. Her head was bent, but she wasn't asleep, and I went over and said her name. Nothing. I crouched down and held her hand in mine. She squeezed my hand back, but still she didn't say anything. Was it the cancer? She didn't seem to be in pain, but was she depressed?

I was concerned about Joan all day. Even as I talked to other patients and sat through meetings, I couldn't stop thinking about her. When I returned to the side room in the afternoon, Joan's sister was there, and before I could get out my giant ramble of a worry, she said: 'It's Joan's hearing aid – it's stopped working.'

Of course it had. It was a junior doctor's schoolboy error – the patient was failing to answer simply because she couldn't hear, and the loss of her hearing aid had trapped Joan in a noiseless, sightless world with only her own thoughts to keep her company.

'The nurses haven't had a chance to get it fixed, they're rushed off their feet.'

They were. The nurses were just as pressured as the doctors, trying to hold a ward together on goodwill and with half the staff they actually needed.

'I'll go,' I said.

I had never been to audiology, but I'd seen the signs many times as I walked the corridors. The offices were up several flights of wooden stairs, which became increasingly narrow the higher you climbed, and when I finally reached the reception desk I was all out of breath and more than a little flustered. I had to talk my way through five people before I found the man who mended the hearing aids.

'And you're a doctor?' he said. 'We don't get any doctors up here.'

I thought back to being a medical student, when I'd changed a patient's urine-soaked sheets because there was no one else available to do it (and surely anybody else would have done the same). *What exactly* said my registrar *is your role here?* Three years later and I obviously still hadn't found an answer to that question.

'My shift ended ages ago,' I said to the audiologist. 'There's a first time for everything?'

I tried to do a little laugh, but he just frowned.

They mended the hearing aid while I waited.

I returned the hearing aid to Joan, and she told me off for fumbling around and not putting it in properly, but then she was back. All smiles. She had found us again. We could go back to our bars of chocolate and story reading.

'I'm so glad you sorted it out,' her sister said, as we walked in the corridor outside. 'They've finally found her a nursing home and it would be so complicated to get it done from there.'

'They have?' I stopped and turned to her. 'When does she leave?'

It had all been arranged very quickly. Joan would be

discharged the next day. It was good news, of course. In the small amount of time she had left, it would be far better for Joan to be settled, to be away from the hospital and the risk of infections and isolation. I just couldn't imagine anyone else being in that side room.

'We're not telling her until just before she leaves. She's always hated change.'

My shift had long since ended. By some quirk of the rota I had the next day off and so, after I left Joan and her sister, I wandered around the hospital tidying up loose ends. Typing a discharge letter. Filling out bloods requests. I decided that I might as well and so save someone else the job tomorrow. I sat on different wards and chatted to nurses. I wandered around A&E to see if there was anything interesting going on. The on-call doctor spotted me a few times and looked at me in a curious way. When I checked my watch, it was almost eight o'clock and I thought that it would make more sense to eat in the canteen rather than cook something when I got home.

I sat with my melamine tray at one of the restaurant tables and stared at a plate of untouched food. What was I even doing there? I usually couldn't wait to finish my jobs and get out of the place, yet I was still here three hours after I should have left, and trying to find another excuse to stay. It took me a moment to realise it, but when I did, it hit me with such clarity that I couldn't believe I hadn't seen it sooner.

It was Joan. I knew that by the time I came back to work on Monday she would be gone and I didn't want to leave her. I didn't want to say goodbye.

I got up from my plastic seat and my melamine tray of untouched food, and I walked to the ward.

An evening ward and a daytime ward have very different personalities, and by the time I arrived, all the visitors had left and the patients were settling down for the night. The drugs trolley was making its way around for the last time. Curtains were closed around beds. Nurses sat at computers, typing out the day into keyboards.

I could hear Joan, even as I walked down the little corridor to her side room. She was complaining about the laces on her shoes and I smiled, because there is nothing more reassuring than hearing a patient grumble about something. Very ill patients never complain.

When I poked my head around the door, one of the healthcare assistants was trying to remove Joan's shoes and she looked up at me and smiled. Joan sat on the edge of the bed with her back to me, which didn't make any difference of course, but did somehow make it easier. I looked at her tiny frame, the sloping shoulders and the whorls of white, ice-cream hair, and I tried to make sure that I could always remember this and keep it safe.

The healthcare assistant moved on to Joan's cardigan and Joan moved on to complaining about buttons, and the HCA and I smiled at each other again.

'Goodbye, Joan,' I whispered, knowing she wouldn't hear. 'Travel well.'

And I turned around and I left.

When I arrived back at work on the Monday she was gone and her name had vanished from the patient list.

I didn't have any reason to visit that ward, because Joan had been an 'outlier' – a patient placed on a different ward in a different part of the hospital, because there were no beds available in the usual places. A couple of weeks later, though, I found myself up there again. A different patient. A different outlier.

I was just leaving when one of the nurses called me back.

'I have something for you,' she said, and she reached into one of the drawers in the nurses' station.

It was a cream envelope with my name written on it.

A card with pink and yellow flowers on the front and 'thank you' printed in gold. Joan's sister had written inside:

Joan thought the world of you, you know

and, beneath the beautiful copperplate script, guided by her sister's hand, and in shaky letters of uncertainty, Joan had managed to write her own name.

I knew I would keep the card forever; even after Joan and her sister had long since disappeared and I, perhaps, sat on the edge of a bed somewhere myself, with my sloping shoulders and whorls of white ice-cream hair, because I would need that card as a reminder.

A reminder that sometimes it is better to worry less about what your job might be and more about the tiny decisions we are able to make that will help someone else's journey become a little bit easier.

A reminder that our role in life isn't always the most obvious one.

Buried

There are two things you always remember about being in hospital. The times you felt the most afraid and unsure, and the times when someone showed you a kindness.

The patient

As well as keeping Joan's thank-you card, I also kept Joan.

She travelled with me each day as I moved around the hospital. She accompanied me on ward rounds and we sat together in X-ray meetings. She ate lunch with me each day in the canteen and drove home at night next to me in the silence. She wasn't alone, either. Others journeyed with me too. The thirty-eight-year-old father with pancreatic cancer. The woman knitting in the corner bed. The dying woman in the side room begging for morphine. The young man in ITU. Children in paediatrics with non-accidental injuries. The terminally ill woman in her twenties, whose identical twin sister visited the ward every day and reminded us all of the devastation of cancer. As each day passed, I collected more and more people, and it was inevitable that the weight of those people would prove too much to bear.

The first thing I noticed was how slow I had become. It felt as though I carried the memory of these people in my legs, because each step took every ounce of energy. Unless it was an emergency, I didn't rush anywhere, and it took me the longest time to walk from one end of the hospital to the other. A porter told me that each time he passed me in the corridors, I was always staring at the floor and I realised he was right. I had stopped looking up.

My brain became slower too. I doubted myself constantly. Whenever questions were raised on a ward round or in teaching, I couldn't process what was being asked and I would always be the last to answer. I stared at my jobs list each day, unable to attempt anything on it because the sense of being overwhelmed was so great, so terrifying, it felt like a form of paralysis. If I did manage to complete something, I would return to check again and again and again that I had done it properly, and, if self-doubt is fed and watered by those around you it will soon begin to grow and flourish. On night shifts, whenever I had a few minutes to spare, I would sit in one of the empty offices and take patient notes down from the shelves. At first, it was only the patients belonging to our team, but it soon became any patient. Any department. It didn't matter, as long as there was a story. I would read through each entry, right back to the referral letter, and ask myself whether I would have spotted something on the X-ray, or had the foresight to order a particular blood test. Did I have the intelligence to prescribe that medication or request a CT scan? Was I smart enough? Was I a good enough doctor? Did I deserve to be there?

I became fixated on the risks of spreading infection, and in between patients I would stand at the sink and scrub until my hands became red raw, but still I wouldn't stop, and soon my knuckles started to crack and bleed. One of the nurses noticed and gave me a tiny little pot of Sudocrem, and that small act of kindness in a sea of despair moved me so much that I went into the sluice room and I sobbed. I barely ate. I hardly slept. I lost vast amounts of weight. My hair was matted and stuck to my skull. I would crawl into bed each night and lie in the darkness, unpicking the day, and each morning I would crawl out of bed, dress in the clothes closest to my hands, and I would walk back into a life I had begun to think of as the worst possible living hell. The rest of the hospital coped. I watched other doctors march through their work with what seemed like the smallest of efforts. I walked only at the edges, treading carefully, and my goal each morning was to somehow arrive at the end of day without losing my mind.

If you continue to walk at the edges, it doesn't take long for others to notice. We are very good at spotting when someone isn't pulling their weight, but less good at asking *why* that weight is no longer being pulled. We are also very good at confronting people about it, and several times, doctors who were younger and less experienced felt authorised to tell me off for not being somewhere they felt I should be, or not doing something they felt I should have done. Perhaps they were right, but none of them felt the need to question why that might be so.

I had a small library of people I could visit whenever I felt dangerously unable to cope. Different nurses dotted around the hospital who were especially kind. The bereavement officer, who was one of the most compassionate people I had ever met. The sister on Ward 4, who once made my day by telling me I would have been a wonderful nurse. The hospital chaplain, who always stopped to speak to me and who had the wisest eyes of all. The chapel was right next door to the doctors' mess and I would sit in there sometimes after a difficult shift, even though I had stopped being able to see any evidence of God at all as I walked those corridors. Perhaps it was the comfort of a silence after all the noise, or perhaps it was the chance that somewhere within that silence there lay the possibility that one day I would find Him again.

I had arrived on the wards filled with joy and enthusiasm, and a desire to be the best doctor I could possibly be. The inadequacies of the system, the lack of funding, the absence of people to provide the necessary care, and the misery and the death and the dying had all whittled away at me until there was nothing of that doctor left. She was gone. Sometimes, she felt so far away, I began to wonder if she'd ever existed in the first place. I'd always walked an extra mile, even if I was never the best, but the job had sent me into such a deep well of despair that it took less energy to wear the coat they had now given me: outsider, troublemaker, clock-watcher. It was so very much easier to become that person than it was to swim against the tide.

I had always looked up to the NHS. I had always seen it as a shelter, a constant, somewhere offering protection and safety to those in need, and yet its quiet and petty cruelties took my sense of loyalty and allegiance, and removed it, piece by piece. When my uncle died, I was required to provide a death certificate in order to take three hours off for the funeral. When heavy overnight snowfall prevented me from driving into work, I was forced to wade into the middle of my road, through knee-deep snow, to take a photograph in order to prove that I wasn't lying; and I quickly found that, despite their name, human resources were not especially resourceful and actually contained very little humanity.

The NHS I loved had turned its back on me. Not only had it allowed me to fall, there were times when it felt as though it had given me a small push in order to get me there.

At the beginning, middle and end of a placement, you are expected to meet with your consultant to discuss your time there. Any issues you may have encountered. Any problems. To review your progress and discuss your learning, and to ensure that your spiritual and emotional well-being remain intact. These meetings are notoriously difficult to arrange, due to both the consultant's heavy workload and the junior doctor's endless list of jobs. You are expected to find someone to 'hold your bleep' for the duration of these meetings and very often the only thing you can think about for the whole time you are in there, is how much work is mounting up in your absence. Some consultants prioritise these meetings, some do not. Very often, the middle meeting is forgotten and glued to the final one in an attempt to preserve the impression that someone was concerned about you for those four months.

In surgery and medicine, these meetings often felt like a piece of theatre. You sit in front of a consultant, someone you barely know, someone who you'd frantically tried to keep up with on a ward round, someone who clearly coped very well in the job in which you are now struggling to survive, and the consultant you barely know asks you if you are okay. You weigh up your options. You see the cursor hover over a white box, ready to tick. You sit ashen-faced, with red-raw hands, exhausted, gaunt, yesterday's sleep still sitting in the corners of your eyes, and with a paper-thin voice you tell him that you're fine, absolutely fine, because 'fine' is so much easier than the alternative. Saying anything else would take too much energy, and you know that the small amount of energy that you possess needs to be saved. If you squander it on telling a complete stranger how you really feel, you are worried you might not have enough left in you to make it to the other side of the day.

'I'm fine,' you say. 'I'm absolutely fine.'

But fine is a word that will slide from your throat, and fine is a word that will bury you.

In my placement in general medicine, I sat in front of a respiratory consultant and I was asked this question. He leaned forward as he asked, and for a moment I thought he was going to rescue me. I thought I had finally been noticed. Instead, he listed my inadequacies one by one, carefully dismantling the very last pieces of my self-belief, reducing me to nothing, and he screamed all of this at me with such violence that his spit landed in my face.

After he'd left, a secretary came rushing from an office three doors down, to check that I was okay. She stood in the doorway and looked at me with a curious tilt of the head.

There was more concern in her face than I had seen from anyone in that hospital for the last twelve months.

'I'm fine,' I said. 'I'm absolutely fine.'

Birthstones

There are times I may have been the doctor I did not want to be. I have to accept that it was me who ignored a patient telling me he did not want to be alive during a busy surgical ward round because I did not know what to do with it, that it was me who ignored the distress of a colleague falling ill. It was me and not someone else. Just me. The doctor I did not want to be. When I ask myself these questions, over and over, and think of the possible answers, the self-compassion I told others to apply to themselves has provided forgiveness for some of my shortcomings.

The consultant

Burnout is an unlikely phrase, because it implies that the effects are loud and obvious, raging like a fire for everyone to see.

Most burnout, however, is quiet and remains unseen. It exists behind a still and mirrored surface, deep, out of reach, unnoticed by everyone – even, sometimes, by the one who is burning. You might catch a glimpse of it if you look carefully. You may say that someone is acting out of character or is unusually short-tempered. You may curse a co-worker for constantly missing deadlines or for being too easily distracted. You may notice that someone is always very early (or always very late) and you may detect that they have a lack of interest in the work in which they once took pride. Or you may notice none of these things. You may walk alongside a raging fire, day after day, without seeing even a trace of its existence, until a time when something happens and the flames can no longer be kept under control.

Gill was a patient in a side room on Ward 8. We were the same age, almost to the day.

As children, we'd watched the same television shows and saved our pocket money to buy the same records. We had grown up with the same posters pinned to our bedroom walls and we knew all the same lyrics to all the same songs. The only difference between Gill and me was that Gill had meta-static breast cancer and I did not.

On the ward round, I would stand by Gill's bed in the side room and write in the notes as the consultant spoke. As I wrote, I would think about all the past birthdays and past Christmases. How we each had marked our lives with the same tape measure, and how we had both assumed the same guarantee. I thought about all the lyrics to the songs, and as I looked across at her, it felt as though I was staring into a mirror. The reflection in that mirror was almost unbearable, but I had to keep looking. I needed to find another difference between us, because, if I didn't, I was worried I would never be able to turn away.

Gill had been in hospital many times over the past few years. With each visit, the entries in her notes became shorter and shorter, and the hope became less and less. We were now at the point where the only outcome of Gill's treatment would be to prolong her life by just a small amount. We were also at the point where the small amount of life she had left would be ruined by the very same medication that delivered her that time. It's a decision many terminally ill patients will inevitably face. Quality or quantity. To live by clocks and watches or to turn away from them and find another measurement. It is not for you or I, or any doctor or nurse, to say what we would or wouldn't do in that situation: it's always up to the patient to make that decision by themselves. When I walked into Gill's side room late one afternoon in November, I knew immediately that a decision had been reached.

I had been called to reinsert a cannula. The old one had failed and needed to be changed. Gill lay in the semi-darkness, exhausted from treatment that made her sick and weak, that made her too drained to lift her head or to bear even the tiniest fraction of light from the window upon her face.

I removed the old cannula and, on a blue sheet, I laid out the equipment for its replacement. Swabs and gauze, syringes, water, tape and bottles. As clinical and sterile as it seems, taking blood or inserting a cannula is such a personal, intimate act, because the first thing you do is to search for a vein, and in order to do that you need to hold the patient's hand. The times I have taken the hand of an elderly, lonely patient and held it in mine to look for a vein, and they have immediately squeezed my hand back in return.

I held Gill's hand and looked at the tired, worn veins for a possible candidate. I even checked the vein just below the thumb, affectionately known as 'The Houseman's Friend'. There was nothing.

'You don't need to find a vein,' she said.

I looked up. Her face was as pale as the pillow, and she was so frail it seemed as though she was slowly disappearing into the bed. There are terms used in cancer, for example sarcopenia and cachexia – the wasting away of lean tissue and muscle mass – but there is also a certain look you can spot in someone who is terminally ill. It doesn't have a text-book name or an official definition, but to someone who has walked the wards for long enough it's unmistakeable. It's a look that tells you this person does not have very long left. Gill had that look.

'I don't want another cannula,' she said. 'I've had enough.'

It was discussed with the consultant and the Macmillan nurse, and with Gill's elderly parents, who were never very far away. It was agreed and decided. There were tears and sadness, but also, it seemed, a strange sense of liberation, as if Gill had finally taken back the reins. Cancer makes so many decisions for you, it must be empowering – even in those distressing circumstances – to make a decision for yourself.

I was on a set of nights that week, and the first thing I did before my shift began was to visit Gill and also her parents, who had set up home in the little side room – her mother on a pull-out bed and her father sleeping in an armchair. Sometimes they had a question about the medication that was keeping Gill comfortable, sometimes I think they just needed the presence of someone else in the room for a while, a reminder of life outside the hospital and a conversation about the mundane and the ordinary. A few minutes of escape. I would see them occasionally during the night, walking out anxiety and cramped joints along the long, silent corridors, and every evening when I arrived for my shift, I would expect to see Gill's name missing from the list of patients. She held on. We waited.

After my run of nights, I had a day before I was due on call. On the rota, that day looked like a day off, but in reality it was just twenty-four hours in which to try to recalibrate and adjust your body clock to being awake in daylight again. Before I headed into A&E, I went to the ward and sat at one of the computers, and I'd just started scrolling through the patients waiting in the emergency department when Gill's mum ran down the corridor to the nurses' station.

'Could you come now,' she said. 'Gill's breathing has gone funny.'

If you have ever sat with someone at the end of their life, you will know what Gill's mum was hearing. Textbooks try to describe this kind of breathing. They give it a name and attempt to analyse it, but you can't imagine its distinctive sound until you witness it for yourself.

I looked at Gill. She lay back with her eyes closed and her face was ironed of all the agony and the pain she had been through. She was more relaxed than I had ever seen her before and, in those moments, I saw a glimpse of what Gill was like before the cancer. When she was just Gill.

'I don't think it will be very long now.' I turned to her parents. They looked small and broken. 'Would you like me to sit with you?'

I didn't think for a moment they would say yes.

'That would be lovely, Jo,' said Gill's mother. 'If you wouldn't mind?'

Of course I didn't mind.

Her parents sat either side of the bed, and I put my bleep on silent and stayed on a plastic chair against the far wall. The windows of the side room looked on to a little path that led to the staff entrance of the hospital, and I could hear footsteps beyond the drawn blinds. Everyday chatter. Laughter. Clocks that hadn't stopped.

Over the top of this was Gill's breathing. Slowing. Leaving.

Gill's mother looked at me. 'I don't know what to talk about,' she said.

'Why don't you talk about Gill before the illness?' I replied. 'Things that made her laugh, adventures you had together, what she was like as a child. Then the last thing she'll hear is happy memories.'

For the next few minutes, I listened to a life lived parallel to mine. Paths never crossing, but knitted together by pocket-money stories and camping holidays, and posters pinned to bedroom walls. And as the stories unfolded, the time between Gill's breaths became longer and longer.

Until.

'Gill hasn't taken a breath for a long time, has she?' her mother said.

'No,' I replied. 'No, she hasn't.'

I walked over to the bed and held my fingertips against Gill's skin. I watched for the rise and fall of her chest, for the slightest flicker of movement. We stood, the three of us, only for a couple of minutes, but it felt a forever time. I knew, even without the checks and the watching, because the air was different. The room had changed.

'I am so sorry, Gill has died,' I said.

The footsteps on the little path outside and the sound of the ward beyond the side-room door all seemed to fade for a moment, and we were held still in the quiet. Perhaps the noise continued to be there, but the weight of the room made it unable to be heard, or perhaps Gill needed the silence in order to leave us. The next sound I was aware of was Gill's mother crying, very softly. The crying of a person who has lost someone they love to this brutal disease – a mixture of relief at an end to their suffering and despair at what might have been. Grief for the disappearance of hope and a lost future.

Her mother stood up and pointed. 'Could you straighten Gill's necklace?' she said. 'It's crooked. She hated her necklace being crooked.'

Gill was wearing a very fine chain with a small amethyst pendant. Her birthstone. My birthstone. I gently reached around her neck and adjusted it so the clasp was at the back and the tiny stone lay flat. It was the kind of thing you'd do for a friend or for your mum.

'I am so sorry,' I said, because I felt the tears overwhelm me. 'I am so sorry.'

There are times when you can force crying back inside you somewhere, when you can make it wait for a while, and there are other times when you feel unable to hold on to that control, when the emotion overpowers you with such force that the only thing you can do is to let go.

The tears weren't just for Gill. They were for the elderly couple who had just watched their only child take her final breaths, they were for the injustice and the misery and the broken system. For all the times I'd forced the crying back inside somewhere.

'It's so unprofessional,' I said. 'I am so sorry.'

Gill's mother put her arms around me. 'You are a human being first and a doctor second, and I can't tell you what a comfort it is knowing how much Gill meant to you.'

There we stood, the three of us in curtained light. All crying for someone who deserved so much more of life.

I left Gill's parents to say goodbye and I walked back on to the ward. The light and the noise and the cruelty of an ordinary day felt like an assault. I was still crying. I wasn't sure I would ever be able to stop.

The other junior doctor took one look at me and lifted the bleep from my hands.

'Go,' she said, because she was kind and understood. 'Go!'

I left the ward and pushed through the doors on to the little path outside Gill's room. I climbed a steep bank, where a shortcut had turned the grass into mud, and I walked to the far side of the car park. My car was the only place I could think of where I could be alone, and I sat in the driver's seat and I sobbed. Huge, angry sobs that made my body fold and retch, and my lungs hungry for air.

I stared over at the hospital and wondered how I had ever imagined I could do this. Perhaps it might have been possible to face the misery and unfairness inside it each day with the right support. Although there were wonderful, kind doctors working inside that building, there were also the ones who passed you in the corridor without a backward glance at your welfare, whose spit landed in your face, the ones who seemed happy to watch you fail. These were the people who walked through your mind in the early hours, who kept you from sleep, who disallowed you the joy and the privilege of your job.

Perhaps, as juniors, they had been bullied too. Perhaps it was a family heirloom they felt obliged to pass down to the next generation, or perhaps not all good doctors are good people.

I knew as I sat there that I couldn't stomach another mouthful, and every ounce of self-preservation told me to start the engine and drive away. Although I didn't know where I could go. I couldn't return home and display my own failure and inadequacy for everyone to see, but there was nowhere else I belonged, and I wondered momentarily whether it would be easier, perhaps, just to disappear. I sat with the engine running for a long time, searching for a thread of something to keep me there. Eventually, I found it. The patients.

Being with Gill's parents was a job I had done well. If I managed to make the tiniest difference, if I helped to make the most traumatic experience of their lives even the slightest bit more bearable, it meant that I was learning to be a good doctor after all. (I didn't know it then, but a week later her parents would thank me in Gill's obituary in the local newspaper.) Gill's death was one of the worst days I experienced as a junior doctor, but it also reinforced my decision to work in psychiatry. As a medic, you are responsible for monitoring your patient's blood pressure and calcium levels and pain relief, but you are also responsible for monitoring their worries, their emotions and their hopes. As part of a patient history, a doctor will always ask – among many other things – about the history of a presenting complaint, about the patient's medication and about their previous illnesses. Tagged on the end, along with drug allergies and smoking status, there is a small section in which you are required to ask about the patient's emotions. In another of medicine's wonderful acronyms, it's known as ICE (ideas, concerns and expectations) and it's often given the least time and has the shortest answers of all. Perhaps it would be better to start the consultation with ideas, concerns and expectations – and not only those of a patient, but also of a patient's family and friends. And perhaps this section of a history should be the one that is given the *greatest* weight and time, because these are the things that shape both our lives and our health. This is the lesson I learned that day, and, having the privilege of meeting Gill and her parents, and waiting with them in that curtained side room, made me appreciate that medicine is so very much more than a science.

I nearly drove away from medicine that day, but eventually I reminded myself that I was committed to this job, and I knew, no matter how I felt, that I had to go back inside. To desert the patients was unthinkable. To walk away from an obligation and a promise wasn't an option, no matter how desperate I was, and so I walked again along the little path and through the double doors and the ward where Gill's parents still sat with their daughter. I walked along corridors, past porters and nurses and medics and their ordinary days, past the clattering of laundry trucks and the wheels of patient trolleys, and through the swing doors of A&E. I took the first patient notes from the desk and I sat in a cubicle, behind paper-thin curtains, listening to the rest of the hospital happen around me. My hands shook and my eyes swam with too much seeing, and I wondered how someone could walk through a landscape and be at the very lowest point of their life and yet no one who passed by them even noticed.

I had two weeks annual leave.

It had sat in my pocket for the longest time, and I had

almost given up being able to take it. Because there were so many gaps on the rota, it was down to us to find someone to cover our shifts if we wanted to take a holiday to which we were entitled, and it was almost impossible. Days were lost, much-needed breaks slipped off people's calendars and disappeared, but, as I worked through my shift in A&E that day, I knew that I had to find some way to escape, even for a short while. If no one was prepared to help me, I had to help myself, and through bribery, promises and sheer determination, I eventually managed to put it in place. I had my fourteen days to recover.

During those fourteen days I didn't go on holiday. I didn't fly anywhere exotic or even book a little cottage by the sea. I just did what I love doing the most. I read. I did nothing but read. From the minute I woke, to the minute I went to bed. Thrillers, classics, poetry. Plays, autobiographies, essays. I filled my mind with other peoples' words and thoughts, as many as I could, and at the end of those fourteen days I had read sixteen books.

I strongly believe in the power of words to heal and mend. We read stories to make sense of the world, to better understand our own situations and challenges. Every story we read follows a template, a prearranged agreement between author and reader that states that, despite many obstacles, the narrator will ultimately succeed on his journey, and the villain will eventually get his just desserts. We expect this. It's part of the deal. When we read the last page of a book, we feel safe in knowing we will be given our happy ending, even if that happy ending isn't always the one we might expect it to be.

We have followed this template since we were children,

since the very first time we were told a bedtime story, and we take it with us into the real world, perhaps subconsciously hoping that the template will work out there too. It doesn't, of course. The obstacles faced by many of my patients weren't surmountable and none of their last pages felt like a happy ending. Perhaps our constant despair at the world, at the politicians, the deep lack of humanity and the many injustices of society, is felt more sharply because it fails to fit into the template we've believed in for so long.

Perhaps this is why reading is so important, because in reading we restore the possibility of hope.

After my two weeks were over, I prepared myself to go back to the wards. This time, however, I wasn't jumping back into the turmoil of general medicine or surgery. Instead, I was heading where I had wanted to work since the first day of medical school, and perhaps even before that – since I delivered pizzas and typed letters, and chased people around department stores. Since before I dared myself to see if I was smart enough to be a doctor.

I was heading to psychiatry.

I had looked forward to this moment all the way through medical school. I had tried to focus on it as I walked through the agonies of my previous rotations. It was the whole reason I was there in the first place. Yet still the misery I had endured in the previous few months lay shallow at the surface and I decided I would give it a week. If, after a week, I felt the misery creeping back, I would unfasten myself from medicine forever. No matter the shame and the humiliation, and no matter the inevitable chorus of 'I told you so'. I would give it a week. One whole week.

I was there almost a year. Until the junior doctors' rotas changed and I was forced to say goodbye.

Psychiatry rescued me.

Minds

I have heard voices since I was a child. I used to think everyone did, until I got older and realised something was wrong. I've been in hospital more times than I can remember, and I've been sectioned on four occasions. I don't remember some of the admissions. Perhaps moments of them, but then it feels like looking at a photograph of yourself and not remembering how it came to be taken.

I always remember when people were nice to me, though. Daft things, really. When someone lent me their phone charger or gave me their chair in the day room, because they knew I liked to sit next to the window. Nurses who listened to me. Doctors who cared. I hold on to those moments when I'm ill, because when your life is filled with bad things, you have to keep polishing the little pieces of good just to keep you going.

The patient

On my first day in psychiatry, after we had done the patient reviews and gone through the drug charts, and I had typed up the discharge letters, I asked how I should spend my afternoon.

I was still smarting from my experience in general medicine and was determined to show I could be a good doctor. I wanted to make a decent impression, to give this job the best shot I could before I admitted defeat; before I finally threw in the towel for good.

After a brief pause, I was told I should talk to the patients.

I had just spent a year being told that I talked to the patients too much.

The relief was unimaginable.

Of all the underfunding in the NHS, it is most noticeable by far in mental health services. Perhaps that's because here you often find people who have nothing to begin with. While patients I had met before usually had a framework of people they could rely on – to make sure they took their medication, to do a bit of shopping for them until they were back on their feet, to stand their corner, or just talk to them – there were mental health patients, thanks to a life spent largely being ignored and excluded, who had no one. Not a soul. I met one woman with schizophrenia who said that she'd stopped taking her medication because it took away the voices she heard, and they were the only ones keeping her company.

When the embroidery of life is taken away, when those who support and bolster and underpin are removed from our picture, and we are standing completely alone, only then does the true effect of isolation become apparent, and the NHS is left to deal with the consequences of those losses. The small spaces where people used to find comfort and companion-ship. The libraries. The coffee clubs, the inner-city projects and the village halls, all cut away and vanished, leaving whole armies of people with nowhere to go. Without these spaces, communities themselves eventually fragment and disintegrate. Whereas once we would notice if someone in our neighbourhood was struggling and unable to cope, now roads and avenues unfold endlessly across a landscape and houses are filled with people we neither meet nor care about.

'Nobody on my street knows my name,' said one patient. 'No one would even notice if I disappeared.'

Some of the patients on a mental health ward have spent their lives drifting unanchored, existing alongside serious illness without support or even acknowledgement, until one day that serious illness causes them to catch and snag on a corner of society, and the rest of the world becomes aware of it. Sometimes, the catch is small and manageable. Sometimes, it is not. Sometimes, the unmanageable is caught just in time, like the accountant who stood on a train platform and heard the voice of God telling him to push a stranger on to the tracks.

'I wanted him to die because he's part of The Club,' the accountant said. 'They're after me. They're everywhere.'

His eyes pleaded for something and his voice carried with it a question, and I wasn't certain if that question was a need for acknowledgement or a need for a release from the person his illness had forced him to become.

Sometimes the unmanageable isn't caught in time and this is when we see hysteria and screaming headlines, as society forgets that the patient and the illness are not as one, and we blame the person when we really should be blaming our own failure to notice when that person is first in need of help.

Others on mental health wards seem to arrive from nowhere, from quiet, anchored lives, from suburban houses on leafy streets where you wouldn't imagine mental illness would wander at all. The first thing you will notice as you walk on to the ward is a gathering of people from all backgrounds and from all circumstances. There are those without homes or jobs, and those with families and professions. Those who have lived with mental illness their whole lives and whose notes span several volumes, and those who find that one day they are suddenly no longer able to cope with whatever life has delivered to them. It is also fair to say that on every mental health ward and in every psychiatry outpatient clinic, you will find a huge army of midwives, doctors, nurses, pharmacists and social workers. You will find the NHS, stretched to its very limits.

The second thing you will notice as you walk on to the ward is that all those people, and all their variety of backgrounds, have become a community. There is often tension and argument, and the dynamic changes with each new admission, but similarities are celebrated and differences are forgotten. Patients support each other. Friendships are made. Perhaps thrown together by chance and circumstance, unity is built out of diversity and the ward supplies what society has so often failed to provide – a sense of belonging.

I could feel it as soon as I arrived on my first day, as soon as I pushed open the double doors and walked on to the ward, and I knew instantly it would provide the very same thing for me too.

So many nurses are extraordinary, but I have never met more extraordinary nurses than I did in mental health.

Not just the nurses, either. The support workers, the occupational therapists, the social workers and the volunteers. The doctors, the ward managers, the pharmacists, and the speech and language therapists. An entire population of people whose only purpose in life is to give a patient back their self-belief, and to rescue a life worth living. So many times, I saw small moments of compassion – so fleeting, so transient, they could easily have gone unnoticed. If I told them here, they would seem insignificant and they would be lessened by the telling, but to watch them across a ward or a day room took my breath away. Because those moments reminded me of the kindness one human being can show to another, to a stranger. This kindness is nothing to do with wearing a uniform or holding a stethoscope. It's to do with being human, and in psychiatry, I began to witness the very best of humanity.

Miracles

Psychiatry is probably one of the most significant two-way processes in healthcare. You can see such a change and it refreshes and re-energises you as much as the change benefits those you work with. The emotion that can break your heart is sometimes the very one that heals it.

The mental health nurse

If medicine is a book of stories, psychiatry holds the wisest chapters and the ones from which you will learn the most.

Within each patient narrative is an opportunity to understand – not just about the illness, but about wisdom, humour, life and people.

'How could you enjoy working in a place like that?' I hear, over and over again. 'Don't you feel unsafe?'

In general medicine and general surgery, I have felt unsafe on many occasions. I have been assaulted several times in A&E.

In psychiatry, I only ever felt unsafe once, a few years after my first psychiatry job, when I was working in a different NHS trust, on a high-dependency ward.

Daniel was diagnosed with schizophrenia when he was

nineteen. Now forty-seven, the previous twenty-eight years had eaten into who Daniel was and who Daniel might have been, if he hadn't been forced to live his life alongside a serious illness.

He was a 'revolving door' patient, someone who was frequently admitted.

'Daniel's back,' one of the nurses would say, and no one had to ask 'Daniel who?'

Daniel was on a 'CTO', a community treatment order, a (contentious) part of the Mental Health Act that allows patients to be discharged from hospital and remain in the community but only if they adhere to certain conditions – taking their medication, for example, and keeping appointments with their community mental health team. If any of the conditions of the CTO are broken, you are immediately brought back into hospital. Daniel was brought back into hospital on numerous occasions. His daily tablets had been swapped for a monthly injection, in an attempt to simplify things both for him and the people looking after him, but Daniel would disappear whenever the injection was due. He drifted between houses, sleeping on sofas, living in the shadows of other people's lives until the community team eventually managed to catch up with him.

On this occasion, Daniel had been brought in by the police. He was agitated and aggressive because he hadn't received any medication for weeks and he was deeply unwell. Afterwards, the two coppers sipped tea in the nurses' office. One of them showed me how the police restrained people by forcing the thumb back towards the wrist. He called it 'soft restraint'. It didn't look very soft to me, and although I could appreciate the occasional need to restrain a person for the safety of others, let alone their own safety, I couldn't imagine how it must have felt for Daniel. Ill. Afraid. Alone.

In a perfect world, Daniel would have been admitted to a PICU (a psychiatric intensive care unit, designed for people who are acutely unwell and with specific circumstances and needs). In our imperfect world, the only PICU beds available were out of area, which would not only be upsetting for Daniel and anyone who wanted to visit him, but would also be extremely costly for the Trust. And so he was placed with us first (one step up from a general ward, but not as well-equipped as a PICU), just to see if it was workable.

It was not.

Daniel was tall, broad, aggressive and loud, and the other patients were afraid of him. Like many psychiatric units across the country, it was a mixed ward, and its demographic was hugely variable. Middle-aged women with bipolar sat next to young men with OCD. Older patients, fragile and uncertain, and suffering from the psychosis of late stage Parkinson's disease, shared their space with men like Daniel, who were often unpredictable and violent.

Daniel's illness made him throw furniture around and pull a door from its hinges. It made him scream at other people and at himself. It made him repeatedly bang his head against the wall in the corridor. Daniel's illness made it necessary to restrain him multiple times. He was taken to 'seclusion' – a safe room from where he was not permitted to leave – and he was injected with medication against his will. I was not part of the team that restrained patients – it involves very specific training and guidelines and is only used in the most extreme and necessary circumstances – but I have witnessed it several times and it's the most disturbing thing you will see in psychiatry. It's the most disturbing thing you will see in any specialty. It is not so much the patients who fight the restraint that makes it disturbing, it is the patients who don't.

Daniel desperately needed a PICU bed and we were in the middle of organising one for him when there was another emergency on the ward, involving a different patient. The specialist team had all rushed to the other side of the unit and I was alone in the office. I finished what I was doing and closed the door behind me. The corridor I walked on to had the locked exit at one end, and in the other direction was the main ward and the patients' communal area. It was deserted. I turned, locked the office door, and I decided to start walking towards the ward. When I looked up, though, the corridor wasn't deserted any more, because in front of me stood Daniel.

I weighed up my options as he stared at me. I could turn left and use my swipe card to leave, but it would run the risk of Daniel following me and I knew he'd be gone. Along the corridor, the only other doors between me and Daniel were for the treatment room and the laundry. Both of those doors were locked. I could go back into the office, but that would mean turning away from Daniel and fumbling with my keys, and my gut told me that wasn't a wise thing to do, and so I walked towards him. I had no choice.

He seemed to fill the corridor. I tried left, right, left, but he blocked me each time.

Instinctively, I reached for the alarm on my belt that would alert other staff to a problem. It wasn't there. The ward didn't have enough to go around and it was decided – under-standably – that the doctors were not at risk as much as the nurses were. My pulse hammered in my throat, but it was important to stay calm.

'Could I get past you, Daniel?' I said, trying to keep my voice as level as possible.

He leaned forward. I could feel his breath on my face.

'No,' he whispered in my ear.

His body blocked my view of the corridor, but I tried to listen for footsteps, or voices, in the hope that someone might be nearby. In the distance, I could hear the rest of the staff dealing with the other emergency. There was no one. Daniel had picked his moment beautifully.

He stepped back.

'I've got something for you,' he said, and he raised his right hand.

In that moment, I wondered how much damage he was going to do. Would he knock me out? Would he be able to fracture my skull? Would he hit once or would he keep striking me? Once I fell, would he start kicking me? How long would it be before someone realised? My legs weakened. I took a deep breath, hoping it might help me deal with whatever was going to happen next.

His hand came towards me with such force and speed, and I shut my eyes against the impact. But there was none. He stopped, just short of my temple and instead of hitting me, he ran his fingertips down the side of my face.

'I've got something for you,' he said.

I heard movement somewhere behind him and when I opened my eyes and looked beyond Daniel's shoulder, there were three more patients standing in the corridor. A little old lady, whose diagnosis changed with each visit to hospital, a young woman who had spent her entire life living with bipolar, and an elderly man who was admitted with depression after his wife died. With one sweep of his hand, Daniel could have knocked them all flying like skittles.

The little old lady took a step forward and jabbed her finger into Daniel's back.

'YOU LEAVE DR JO ALONE!' she shouted. All five feet of her.

He lifted his hand from my face. He turned around and stared at the little old lady, and after one brief moment of hesitation, he did exactly as he was told.

I think he was so shocked that he went back to his room like a naughty schoolboy.

The little old lady turned to me. 'Are you okay, love? Do you want me to make you a cup of tea?'

The kindness of patients is everywhere. Studies show that an act of kindness not only benefits those who receive it, and provides a sense of well-being to those who give it, but even those who watch from the sidelines feel better just by witnessing it.

It is a true saying that those who have the least give the most. I have seen patients share their very few possessions and clothes with someone who has been admitted with nothing. There are some people who never have visitors or anyone to care about them, and during visiting hours, I have witnessed one patient invite another to join their family instead of sitting with no one. I have seen those who have been on the wards a long time make a cup of tea for someone who has just been admitted, afraid and alone. When you are world-weary, or ward-weary, when you have had your fill of unkindness and cruelty and suffering, to witness small and quiet acts of compassion restores your faith in the world like nothing else.

A PICU bed was found for Daniel and he spent two months there. Once his treatment began to work and his symptoms were more under control he came back to us, and the first thing he did when he arrived on the ward was to walk up to me and apologise.

I had never experienced physical intimidation before Daniel, but I had been verbally abused many times by mental health patients who were unwell, because when you are ill and afraid, when you feel trapped and helpless, you will take any weapon you can to defend yourself. In every single case, without exception, when the patient recovered they apologised to me – although none of them, including Daniel, had anything to apologise for. Their words and behaviour were the symptoms of an illness, in the same way that any physical illness gives us symptoms that are beyond our control.

As a society, we disregard the symptoms of mental illness and we view them as the person and not the disease. The language we use further dilutes them, until they become lost in the mundane and the everyday. OCD is not going back to check that a door is locked: OCD is walking along the middle of a dual carriageway picking up litter, because its presence brings an anxiety you are unable to bear. OCD is not being particular about the way your cupboard is arranged: OCD is urinating in your front room, because the rituals and counting exercises you are forced to complete before you are allowed to walk to the bathroom are so complex and so time-consuming that they do not allow you to get there in time. Schizophrenia is not a 'split personality': schizophrenia is sprinkling flour on the treads of your staircase because the voices you hear are so real that you want to catch the person who must be hiding in your house. Depressed is not a reaction to your football team losing: depressed is being consumed by a despair and a self-loathing that is so overwhelming you would rather end your life than continue to carry it with you for a moment longer.

To remain standing under the weight of these illnesses is a sign of the most enormous courage. To retain your humanity and kindness to others under that weight is nothing less than a miracle.

Peripheries

Visiting time on the medical and surgical wards is always chaotic.

There are never enough plastic chairs. Families crowd around beds, despite the rules. Relatives (understandably) hunt down doctors for information. There is no point trying to do anything for a patient during visiting hours, because you would often have to wade through a vast sea of people in order to do it.

Visiting time on a mental health ward sometimes passes unnoticed. There are, of course, patients with incredibly supportive friends and families – support that plays a huge part in helping recovery – but for many people, visitors are few, and often completely absent. Occasionally, someone wants to keep their admission a secret, because the stigma attached to a psychiatric admission, sadly, has many and long-lasting repercussions. Usually, though, this happens because the patient has spent a lifetime alone. Families have broken apart, friends have drifted away. Here, often, are people who live on the periphery, people who are never included and rarely acknowledged. In every town, in every village, even on your own street, there will be someone who is isolated and ignored. Chances are, they are also suffering from a mental illness.

It's difficult to imagine how that exclusion might feel, but if you work in psychiatry, you will occasionally see a glimpse of it.

A few years after my first experience of psychiatry, I was working on a different ward for a different NHS trust. I had only been there a couple of days and, in the confusion of a new routine, I left my swipe card and lanyard at home. Because the unit I was working in was locked, it meant, for that day, I had to rely on other staff members to let me on and off the ward. It was a nuisance. So much of a nuisance, I knew I wouldn't forget my swipe card again.

I had just left the ward to collect some patient notes from one of the secretaries when I found myself a short distance behind a social worker in one of the long corridors. I knew her from a different job, many months ago. We'd only met once, but she had very distinctive red hair, and I recognised her straight away. She also happened to look after one of my favourite patients. We were the only people in the corridor.

'Hello!' I said. 'How is Leo doing?'

She was only a couple of steps in front and she turned. There was no reply, she just glanced at me, up and down, and then she turned back and carried on walking.

I was puzzled. She definitely heard me. She even looked at me. 'You take care of Leo, don't you? I just wondered if he was okay?'

She kept walking, quickened her step. I quickened my step too.

We turned into another empty corridor.

'Excuse me,' I said a little louder. 'How is Leo?'

Still I was ignored. If anything, she walked faster, an occasional trot in between the steps.

It was baffling. I came to a halt. Gave up.

'It's Doctor Cannon!' I shouted, in a last-ditch attempt.

Finally, she stopped. She turned around and walked towards me.

'I am *so* sorry.' She gestured to her neck, where my lanyard would normally sit. 'I thought you were a patient.'

The many quiet acts of cruelty directed at mental health patients must accumulate. Before I started working in psychiatry, I had spent a brief few months on the wards feeling as though I didn't belong, and I had the smallest taste of what it means to not fit in. I still had a home and a family, I still belonged somewhere, but even then I wasn't able to cope. To spend your entire life feeling that way, with no shelter, no respite, is unimaginable.

A few months after I had chased a red-headed social worker down a corridor, I was standing at the nurses' station on the ward, talking to some of the other staff. There were quite a few of us – nurses, a pharmacist, support workers. I liked being at the nurses' station, rather than shut away in an office, because it's impossible to learn anything if you don't spend time on the shop floor.

It was almost break time and we were talking about nonsense – food and holidays and television programmes. One of the patients came over and joined us. Rob had been on the ward many weeks. When he first arrived, he was agitated and paranoid. He was convinced that the ward was filled with cameras and he was being watched. He thought we were all working for the government and a chip had been placed in his ear to monitor all his activities. Each day he would beg me to remove the chip and set him free. With a change in medication and the right support, Rob had slowly improved. There is no greater privilege than to witness the symptoms of an illness fade and to get to know the person who waited beneath them. Rob was a wonderful man. He lived on a canal boat with his dog, and he loved art and poetry. He knew more about nature and the countryside than anyone you will ever meet. He reminded me a little of my dad.

Rob and the nurses and I were laughing about something we'd all watched on TV the previous evening when the ward clerk came out of the office. She was new, very pleasant and deeply efficient. It was break time and she had decided to make us all a drink, but – being new – she needed to count the teas and the coffees, the milks and the sugars. She systematically went down the line of people standing at the nurses' station, asking us all what we would like. When she reached Rob she just skipped over him and asked the next person. As if Rob wasn't there. As if he was completely invisible. I caught the eye of one of the healthcare assistants and we both stared at each other.

It wasn't the ward clerk's fault. She was new. She wasn't supposed to make tea for the patients. But still.

In the end, I made Rob a cup of tea myself. It felt like the only thing I could do to make the situation a tiny bit less painful. When I handed it to him, he smiled at me.

'Don't worry, Dr Jo,' he said. 'It happens all the time.'

Landscapes

Being a wild card isn't always easy. You are occasionally mistaken for someone else (*are you a social worker?*) or for someone with more knowledge (*are you the lecturer?*) and very often, people think you have far, far more experience than you actually possess (*I want Dr Cannon to take my blood* – no, no you really don't).

Wild cards carry with them uncertainty, doubt, ambiguity. Why are you here now? Where were you before? Why didn't you arrive at this point sooner? A wild card feels a constant need to explain and justify themselves, and there is a certain comedy value in being the junior doctor on a team where everyone else is a very great deal younger than you are. Sometimes, though, what you did before, and the fact that you didn't arrive at this point sooner can, strangely, be played to your advantage.

In my first rotation in psychiatry, in my newly qualified role, I sat in morning ward round waiting to discuss the previous day's events. I had already spent four months there, and I was the most comfortable I had ever been in my life. Psychiatry was why I had applied to study medicine in the first place. It was all I had ever wanted to do. The thought of sitting in that seat had got me through five years of medical school and the challenges of my previous rotations in medicine and surgery. The team was amazing, the patients were brilliant, and I was welcomed into what immediately felt like a family. Other wards can sometimes seem fragmented and there is a strange disconnection between different departments, but in psychiatry, we were a team and everyone was encouraged to use their individual skills and strengths.

There were people from many different backgrounds: those who had worked in psychiatry for decades, and those who had been drawn to it recently, often due to personal experience or the experience of friends and family around them; those who had moved from completely different jobs, and those who had always wanted to work in mental health. Each of us was listened to, every opinion was valuable, and it was the first time I was asked what I actually thought about something. It was such a tight jigsaw of people that it felt at times as if we had all been conjured there by fate and good fortune, and it seemed to me as though everything I had done before – all the pints I'd pulled and the letters I'd typed, all the innocent bystanders I'd chased around department stores – had given me communication skills and an understanding of people you could never teach to someone in a lecture theatre. Everything had happened for a reason, after all. I could finally put the life I'd had to good use.

I was fortunate enough to do two back-to-back rotations in the same job, but that morning happened to be change-over day for everyone else, and a new doctor walked into the handover. I peered at him over my coffee as he was introduced to us.

'This is Dr Smith,' said the consultant. 'Dr Smith is a

Foundation Year 2 doctor. He has more experience than Dr Jo.'

I tapped my finger on the arm of my chair and a bristle of irritation wandered around the back of my neck. Dr Smith smiled at the room. He even did a small bow. He was a good ten years younger than me. He wore a crisp white shirt and a tightly knotted tie, and around his neck was an expensive and very shiny new stethoscope. I thought of the patients next door. I wondered how this was going to work out.

In our first lecture on the first day of medical school, as we were welcomed into our medical career, we were also told something else – something I would perhaps argue about now, but something which broadly makes sense. We were told that there are two kinds of doctor: white coats and cardigans. Those who love the science and those who love the people. Those who order tests for the patients and those who talk to them. Using those (debatable) parameters, I was so much of a cardigan, I was off the scale. Dr Smith, however, was born and bred in a white coat. He had travelled from A-levels to medical school and into being a doctor without taking a breath. I was the wild card. I had taken so many breaths, I was surprised there were any left to go round. Still, the consultant was right. It was true, Dr Smith was more experienced. He was a Foundation Year 2 doctor. He was a year ahead of me on this journey, after all.

We muddled along. If there were new people to clerk, I would take the histories and Dr Smith would take the bloods and the ECGs. In the afternoons I sat in the day room, chatting to the patients, while Dr Smith sat in the doctor's office, working on his audit. Occasionally, he would appear and hesitate for a while at the edge of the room.

'Why don't you join in?' I'd say, later.

'I don't know what to say to them.'

'Them?'

'The patients.'

'Just have a regular conversation.'

He frowned.

'Talk about the same things you'd talk about to anyone else,' I said.

He still frowned.

The patients were a mystery to him. The only problem was that the patients soon cottoned on to this. They invented physical ailments for him to investigate, only to give him a psychiatric history when he tried to examine them. They made fun of his stethoscope. They regularly coaxed out his awkwardness and used it for entertainment. I became tired of rescuing him, partly because he constantly built his own gallows but also because there was a strange sense of satisfaction in watching someone else play the wild card for a change. Shamefully, the more Dr Smith floundered, the more secure I felt in my own foundations. Besides, he should be able to rescue himself quite easily. He was, after all, I told myself, far more experienced than I was.

A few weeks after Dr Smith's arrival, we were both assigned to a new patient as a case study.

She was a young woman, with no history of mental health problems and no prior engagement with mental health services. She had previously been very quiet and reserved. A hard worker. Few friends. Living an unremarkable and un-extraordinary life. However, during the course of one chaotic weekend, and completely out of character, she had stolen a car and driven many miles (without a driving licence) to an unconnected town in the north of England, where she began screaming and shouting at people walking around the shopping centre and threatening violence to anyone who approached her. She was brought in by the police. She refused to speak to any of us.

Dr Smith and I were quite puzzled. I tried to talk to her, but each time I did, she would just walk in the opposite direction. Dr Smith didn't even make an attempt. She didn't speak to any of the staff, or to any other patients, and for the most part she would sit in her room staring silently into the walls. Her parents said that, in retrospect, she had become more withdrawn in recent weeks but that there had been no trigger, no inciting incident. There was no trail of breadcrumbs for us to follow.

Her parents visited each day and we relied on them to piece together a narrative. She refused to speak to them, sometimes sitting at a different table, at other times staring beyond them and into the gardens. Still they came. They brought gifts and food and trinkets from home to make her feel more comfortable.

'Your parents are so lovely,' I said to her one day, as Dr Smith and I walked back to the ward with her after visiting hours.

I didn't expect a reply, but she turned to me. Since her arrival, it was the first time she had even acknowledged that anyone had spoken to her.

'My parents aren't really like that,' she said very firmly.

I glanced at Dr Smith.

'Yes, but don't you think it's a strange thing to say?'

Dr Smith and I were sitting in the office a few minutes later.

'Not really,' he said.

'*My parents aren't really like that*,' I repeated. 'It's just an odd way of describing it.'

'She probably meant they're putting an act on because we're there.'

'But that's not how it sounded,' I said.

The next day, I marched into ward round. I'd spent the previous evening wading through textbooks looking for answers, and I believed I had found one. I was explaining myself even before I'd taken off my coat.

'I know what's wrong with her,' I said, struggling with a sleeve. 'I've worked it out!'

My consultant raised an eyebrow. Even Dr Smith raised an eyebrow.

I explained the conversation we'd had the previous day, the way she'd talked about her parents. The strange words she'd used. 'I think she has Capgras Syndrome!' I said.

My words of triumph disappeared into a silent room.

Capgras Syndrome is a delusion whereby a person thinks that someone close to them – their spouse, their parent, their child – has been replaced by someone else. Someone who looks and sounds exactly like that person, but who is, in fact, an interloper, an imposter.

Capgras Syndrome is very rare.

'I never thought I'd say this to anyone, but I think you've been reading too many textbooks,' said my consultant, and I could see Dr Smith smirk ever so slightly, 'but I'll talk to her.'

He talked to her and it turned out that she *did* believe

her parents weren't really her parents. She thought they were actors, manipulators, fraudsters. She knew for a fact that these people weren't really who they were pretending to be, and – of course – the next step from realising they were impersonators was to destroy them. She was quite willing to discuss this, quite happy to talk to us, it was just that we hadn't asked her the right questions before now, and as luck would have it, I happened to hit upon the right question in that corridor.

'And you knew all this from that one sentence?' said Dr Smith later, when we were writing up the notes.

Perhaps it was luck or just good fortune that I happened to pick up on what the patient was really thinking, but I like to think that the more you listen, the more you hear. If you hear enough stories you will come to know where a beat is missing, where a pause takes the place of a word. You don't have to be a doctor to hear those stories. You'll stumble across them as you're pulling a pint or waiting a table. You'll find them in conversations in a department store and in the queue in a supermarket. The more stories you hear, the more you realise that people always choose their words with care, and words are chosen for a reason. It is, perhaps, something that you learn only with experience.

At the end of the rotation, Dr Smith and I went our separate ways. I went on to another job in psychiatry and Dr Smith disappeared beyond the horizon. If you are worried about him, please don't be: he found his niche, as I had found mine. A year or so later, we crossed paths again in A&E. He was still wearing his crisp white shirt and his tightly knotted tie, and around his neck the stethoscope still looked new and shiny. He was working with the orthopaedic team. He smiled at me as we passed each other.

Dr Smith was no longer a wild card. He looked the most comfortable he had ever been in his life.

Mending

For all the breaking and mending of this profession, it has allowed me to live a life worthwhile. For all the breaking and mending, it has helped me to edge closer to the doctor I wanted to be, the doctor that I still want to become. Because there will always be a need to learn, to improve and to change, very much as I was told on that first day at medical school.

Most important are the words I heard in the past from many people I respected, often on retiring from professional practice, and the same words were very recently echoed by a new friend: let's remember to check in with colleagues that they are okay – because this is what communities do.

The consultant

Many times during my training I would say to myself, and to anyone who would listen, that I wished I had never done this. I would wish that I had taken a different path, that I'd never seen the postcard in the newsagent's window and I'd never persuaded a doubtful professor that I would make a good doctor. I would go to great lengths to list all the other jobs I could have done instead. The many careers that would not have left me mentally, physically and financially exhausted.

Now, I look back and I can't imagine having done

anything else. I can't imagine not meeting the incredible people I have had the privilege to meet, and I am amazed at the fragile decisions I made that allowed our paths to cross, even if it was only for the briefest of times.

One of my deepest concerns about writing this book was that it would deter people from following a career in medicine. If it's any help to those who might be considering becoming a doctor – I truly wouldn't change a thing, even the darkest days, even the days that made me question whether my existence as both a doctor and a human being was really worth anything at all. Because of all the days I had, those were the ones that taught me the most.

Mending, like breaking, can happen in the unlikeliest of places.

It can grow from the briefest event and from the shortest encounter. Breaking is accumulative. We collect small episodes of despair and unhappiness, our own Kodak moments, and we carry them with us until their weight becomes too much to bear and we fracture under the burden. Mending is exactly the same. The more often we witness small moments of compassion, the more humanity we see; and the more likely we are to be able to mend ourselves and the quicker we are to heal.

At medical school, we had many lectures devoted to mending: the anatomical, physiological and biochemical methods of healing, of bone and skin and tissue, from infection and fracture and disease. We learned about the complex process of clotting and the coagulation cascade, the direct and indirect wisdom of bones, the acute inflammatory response and the intricate expression of many thousands of genes. The body's ability to regenerate is remarkable and extraordinary, but it is also intensely fragile and the most important factor ensuring its success is the correct environment. Without the right landscape, none of this can happen. In the wrong surroundings, our body itself becomes a wild card.

When I look at the lives of the people I met in that lecture theatre on the first day of medical school I see many different landscapes. I see surgeons and GPs, anaesthetists and paediatricians. I see those who travelled halfway around the world to follow their career and those who stayed in the hospital where we walked as medical students. I see those who remain working within the NHS and I see those who left and continue to use their skills to help people elsewhere. I agree with many things we were told in that inaugural lecture – it really was the first day of our medical career – but I disagree that there are only two kinds of doctors. I think there are as many doctors as there are people, and as many different landscapes as there are ways of healing.

Psychiatry became my landscape, and within that landscape I learned many things. I learned about the compassion one human being can show to another and I learned about the resilience of the human spirit. I learned that our roles in life are many and valuable and I learned about healing. I learned about the need to look out for each other. I learned about the importance of wild cards.

In a game of cards, the wild card has no suit or colour. It possesses no value. Its currency and worth is determined only by the person who holds it. Wild cards are defined by their landscape and by their keeper, and what is high in value to one player, may be worthless to another. What appears to be a wild card on the surface may, in reality, be anything but.

I am often asked about going to medical school at a later age, about being different, about being a wild card. I always say that the world – and especially medicine – needs more wild cards, but perhaps, if we look more closely, we will discover that we are all different, we are all wild cards. Perhaps each of us is just searching for the right landscape and for our somewhere to belong, searching for the right place to tell our stories, in the hope that someone out there will listen and we will be understood.

Author's Note

My reactions and experiences in this book are based on real events, but the events themselves and the individuals and places involved have been changed to protect the identity of both staff and patients. Details of situations and the people I have met and cared for have been merged and altered to further protect privacy and confidentiality.

Any similarities to particular individuals or events are both coincidental and unintentional.

Acknowledgements

Without my agent Susan Armstrong and my editor Francesca Barrie there would be no *Breaking and Mending*. A very big thank you to both of them, for their kindness, patience and wisdom, and to everyone at C&W and at Profile Books and The Wellcome Collection for believing in my story.

All the thoughts and opinions in this book are my own, but many people have shaped those thoughts and opinions and have helped to make walking this road so much easier. With thanks to the University of Leicester Medical School for believing in a wild card, and with special thanks to Professor Stewart Petersen, Dr Jonathan Hales, Dr Mark McCartney, Dr Tony Dux and Dr Amanda Jeffery. Your teaching and encouragement made stumbling through the back of the wardrobe and into Narnia an infinitely smoother process. To Dr Amy Adams and Dr Cate Bud, who made Narnia a very much nicer place to be. To Dr Chloe Spence, who always understood.

With huge and grateful thanks to Professor Wendy Burn, Dr Kate Lovett, Dr Regi Alexander and everyone at the Royal College of Psychiatrists for your incredible support and for allowing me to be a part of something I thought I would only ever dream about.

To all the incredible NHS staff I have had the privilege to work with, and, always, to the George Bryan Centre for rescuing me.

To all the friends and colleagues who have given their time and words to these pages. To Dr Claire Barkley for answering my many questions on psychiatry and junior doctor welfare, and for providing one of the most interesting conversations I have ever had. To Dr Ignasi Agell for his words and his guidance, and for being the doctor I would like to have become.

Most of all, to the patients. You will always walk with me.

If you have been affected by any of the issues raised in this book, you might find the following resources helpful:

Samaritans
Provides confidential emotional support for people who are experiencing feelings of distress or despair.
Phone: 116 123 (24 hours a day, seven days a week)
www.samaritans.org

Mind
Provides advice and support on a range of topics including types of mental health problem, legislation and details of local help and support in England and Wales.
Phone: 0300 123 3393 (weekdays 9am–6pm)
www.mind.org.uk

Nightline Association
A listening service open at night run by students to support students.
www.nightline.ac.uk/want-to-talk/

Pancreatic Cancer Action
Aims to improve survival rates and quality of life for patients.
pancreaticcanceraction.org

FOR DOCTORS

The Doctors' Support Network
Peer support for doctors and medical students with mental health concerns.
www.dsn.org.uk

BMA Wellbeing Support
24-hour counselling available to any UK doctor.
www.bma.org.uk/advice/work-life-support/your-wellbeing

Practitioner Health Programme
24-hour text based crisis support available for any doctor in England.
https://php.nhs.uk/

**wellcome
collection**

WELLCOME COLLECTION is a free museum and library that aims to challenge how we think and feel about health. Inspired by the medical objects and curiosities collected by Henry Wellcome, it connects science, medicine, life and art. Wellcome Collection exhibitions, events and books explore a diverse range of subjects, including consciousness, forensic medicine, emotions, sexology, identity and death.

Wellcome Collection is part of Wellcome, a global charitable foundation that exists to improve health for everyone by helping great ideas to thrive, funding over 14,000 researchers and projects in more than seventy countries.

wellcomecollection.org